# COMPETITIVE ELECTRICITY MARKETS

# COMPETITIVE ELECTRICITY MARKETS

WLADYSLAW MIELCZARSKI
AND GRAZYNA MICHALIK
EDITORS

**Nova Science Publishers, Inc.**
*Commack, New York*

**Editorial Production:** Susan Boriotti
**Office Manager:** Annette Hellinger
**Graphics:** Frank Grucci and John T'Lustachowski
**Information Editor:** Tatiana Shohov
**Book Production:** Donna Dennis, Patrick Davin, Christine Mathosian, Tammy Sauter and Diane Sharp
**Circulation:** Maryanne Schmidt
**Marketing/Sales:** Cathy DeGregory

*Library of Congress Cataloging-in-Publication Data*

Competitive electricity markets / edited by W. Mielczarski and G. Michalik.
p. cm.
Includes bibliographical references and index.
ISBN 1-56072-603-2
1. Electrical utilities--Australia--Victoria. 2. Energy conservation--Australia--Victoria. 3. Competition--Australia--Victoria. I. Mielczarski, Wladyslaw, 1947- . II. Michalik, G.
HD9685.A83V518 1998 98-35198
333.793'2'09945--dc21 CIP


6080 Jericho Turnpike, Suite 207
Commack, New York 11725
Tele. 516-499-3103 Fax 516-499-3146
e-mail: Novascience@earthlink.net
e-mail: Novascil@aol.com
Web Site: http://www.nexusworld.com/nova



*Printed in the United States of America*

# CONTENTS

# PREFACE

Open electricity, competitive electricity markets are characterised by two main features. First, everyone who is able to satisfy technical and financial obligations has access to generation and retail of electrical energy. Second, electrical energy as a product is separated commercially from transmission and distribution as a service.

Reforms to the electricity supply industry have been introduced in several countries. Victoria, the Australian state, introduced in 1994 the most advanced reform to its electricity supply industry. This has resulted in the competitive wholesale market, where independent private producers offer electrical energy to the pool in bidding offers determining amount of electricity and its price in 0.5-hour trading intervals.

The energy purchased from the pool is offered in the competitive retail market to contestable customers. From July 1998, customers consuming more than 160 MWh per year (that is customers with monthly electricity bills around $1500) will have rights to select a retailer. In January 2001, all customers including households will become contestable customers having rights to select the most competitive retailer.

The Victorian reforms have proved that a new structure of open, competitive markets can deliver low cost energy preserving high reliability of supply and increasing quality of services. The new market is more customer-oriented offering many services unknown under electricity supply monopoly. The success of Victorian market was a base for founding the National Electricity Market (NEM) which the first stage, known as NEM1, started in May 1997.

The articles presented in this volume were prepared in February and March 1998. They describe the electricity market as it was at the end of 1997 and in the beginning of 1998. Over that time, the Victorian electricity market ended its four-years' operation, while the National Market just started to emerge. Therefore, the presentation of the competitive market is mostly based on the Victorian experience with some links to the National Electricity Market.

This Special Issue is organized in five main Chapters:

1. Introduction
2. Market Regulation
3. Trading electrical energy
4. Electricity contracts and price
5. Towards the National Electricity Markets

Of sixteen papers presented in the volume, eight were prepared by the Monash University researchers and students, while the other eight papers were written by economists and engineers participating in competitive market development, regulation, and daily operation. Such a structure of the volume allows the balance between the university and industry presentation.

The first part, Introduction (comprising three papers) presents a historical background of the market introduction, its structure and operation, describing in detail the wholesale electricity market.

The second part, Market Regulation, deals with the electricity market regulations, presenting Codes and Rules, the Office of the Regulator-General and its role, the Electricity Industry Ombudsmen Victoria and her experience, and the perspective of the organizations established to assist market participants and energy users in dealing with competitive market issues.

The third part of the volume, Trading Electrical Energy, presents the wholesale market participants' experience: electricity generator (Southern Hydro) and two distribution/retail businesses. The market managing company – the Victorian Power Exchange, presents VISION, a computer package, which allows the modelling and simulation of the competitive market,.

The fourth part deals with electricity contracts and prices showing categories of contracts, electricity tariffs, and network charges. Demand for electrical energy and statistical analysis of spot prices for nearly four years of the competitive market operation provides the vital information to everyone who may have interest is such a market. Electricity Futures, started in September 1997, represent a new form of hedge contracts leading to market prices stabilization.

A paper presenting the introduction of the National Electricity Market, its structure and operation, concludes the Special Issue.

The volume design is based on the Editors' research on competitive electricity markets and his experience as a technical adviser to the Office to the Regulator-General. The volume aims at the as most comprehensive presentation of the competitive electricity markets as possible in the limited space.

A significant part of the volume is occupied by papers dealing with electricity costs and energy prices. All costs and prices in the volume are given in the Australian dollars. The value of the Australian dollar in relation to American dollar is varying. In the beginning of 1997, it was 78 US cents for one Australian dollar, while in April 1998 this value oscillates around 66 ÷ 68 US cents. Therefore, a long-term exchange ration can be averaged around 70 US cents.

I would like to thank those who have contributed to the volume. The primary thanks must be to the people involved in the electricity market who sacrificed their time writing the volume papers: Dr. John Tamblyn, Regulator General; Ms. Fiona McLeod, Electricity Industry Ombudsman, Victoria; Mr. Roman Domanski, The Executive Director of Energy User's Group; Mr. Grahame W. Kipp, Acting Manager, Power System Evaluation Development and Mr. Gnanbandu G. Gamage, Senior Analyst,

Victorian Power Exchange; Mr. Chrys Chandraraj, Market Analyst, Southern Hydro; Mr. Peter Smith, Acting General Manager, Energy, Eastern Energy; Mr. Steve Abbott, Senior Energy Trader, Australian Gas Light Company; and Mr. Sohail Hassnie, Manager Pricing, the Office of the Regulator-General.

I would like to thank Ms. Grazyna Michalik of Monash University, the co-editor of this volume, whose research experience allowed the preparation of papers on market operation, demand for electricity, and spot price analysis. I would like also to express my acknowledgment to Paul Mielczarski of Monash University, who has prepared a paper on the Electricity Futures in Australia.

Wladyslaw Mielczarski
Editor
Monash University
March 1998, Melbourne

# PART 1 INTRODUCTION

****** 1 ******

# INTRODUCTION TO THE OPEN ELECTRICITY MARKET IN VICTORIA

**WLADYSLAW MIELCZARSKI AND GRAZYNA MICHALIK**
Monash University, Department of Electrical and Computer Systems Engineering
Melbourne, Australia, Email: Wlad@basil.eng.monash.edu.au

**Summary:** In 1998, Victoria is leading Australia in reforms of the electricity supply industry. The reform model is setting the trend not only in Australia, but also in many other parts of the world. The electricity supply industry reforms are making the Victorian economy more competitive giving customers better service, better prices, and better protection. Victoria is not only committed to creating new sustainable markets within the State, but also participates in establishing new national markets. The paper discusses the introduction of the competitive markets in production, transmission and distribution of electrical energy.

## INTRODUCTION

Victoria, one of six Australian States, is located in the south-east corner of the Australian continent. Victoria is populated by about 4.5 million people with over 3 millions living in the state capital city – Melbourne. The Australian administration consists of the Federal Government, located in Canberra, and the State Governments, located in the capital cities of the particular states. The State Governments are responsible for supply of electricity, water and gas.

By 1993, the State Electricity Commission of Victoria (SECV), which was the vertically integrated electricity monopoly, was carrying almost $10 billion of debt, and 30% of its revenue was swallowed up by finance charges. Electricity prices were 40% above the most efficient Australian State (Queensland) and well short of world's best practice, despite Victoria having an abundance of relatively low-cost brown coal and gas reserves as a natural resource. Construction costs of new plant were 60% higher than would be expected under best practice, and generation availability was around 65% compared to current availability standards in excess of 90% [1].

## REFORM OBJECTIVE

The key objectives of the electricity supply industry reforms are to [1]:

- provide the lowest sustainable electricity prices to consumers;
- create the most efficient electricity industry in Australia - aim to achieve world's best practice;
- maximise customer choice and encourage efficient investment;
- ensure an appropriate regulatory environment exists to protect customer service and safety standards;
- ensure long-term security of supply; and
- provide a framework for an effective and sustainable competitive energy market.

The State Government of Victoria is continuing to pursue these objectives, and as the privatisation of remaining State-owned assets takes place, the State will be further removed from commercial risk. The growing National Electricity Market (NEM) increasingly offers a better deal for consumers and places the Victorian industry on a competitive footing.

## 3 OPEN ELECTRICITY MARKET MODEL

### 3.1 FROM THE MONOPOLY TO THE OPEN MARKET

Four characteristic models can be used to describe the structure of electricity supply industry [5]]:

Model 1 – describes monopoly at all levels. A single monopoly company controls the production, transmission and distribution of electrical energy.

Model 2 – allows a single organization, sometimes called the purchasing agency, to chose from a number of different generators. This organization has a monopoly over transmission and distribution network.

Model 3 – allows the wholesale competition. Distribution companies can purchase electrical energy from the pool or directly from energy producers. There is an open access to transmission network; however, monopoly still exists over distribution network.

Model 4 – allows both wholesale and retail competitions. There is an open access to transmission and distribution networks. Distribution of electrical energy as product delivery is separated from the retail activity.

Five years ago, the Victorian supply system was controlled by a single government agency, called the State Electricity Commission of Victoria (SECV). This organization represented the full monopoly from the production of electrical energy to the delivery to individual customers, see Model A in Fig1.

The introduction of the reform to the Victorian electricity supply industry has lead to wholesale and retail competition, see Model B in Fig. 1.

It is a clear analogy between four generic models [5] and two models represented in Fig. 1. Victoria has moved in a very short period from the full monopoly (Model A = Model 1) to competition in wholesale and retail markets (Model B = Model 4). After the year 2000, all energy users in Victoria, including domestic customers, will have rights to select an electricity retailer.

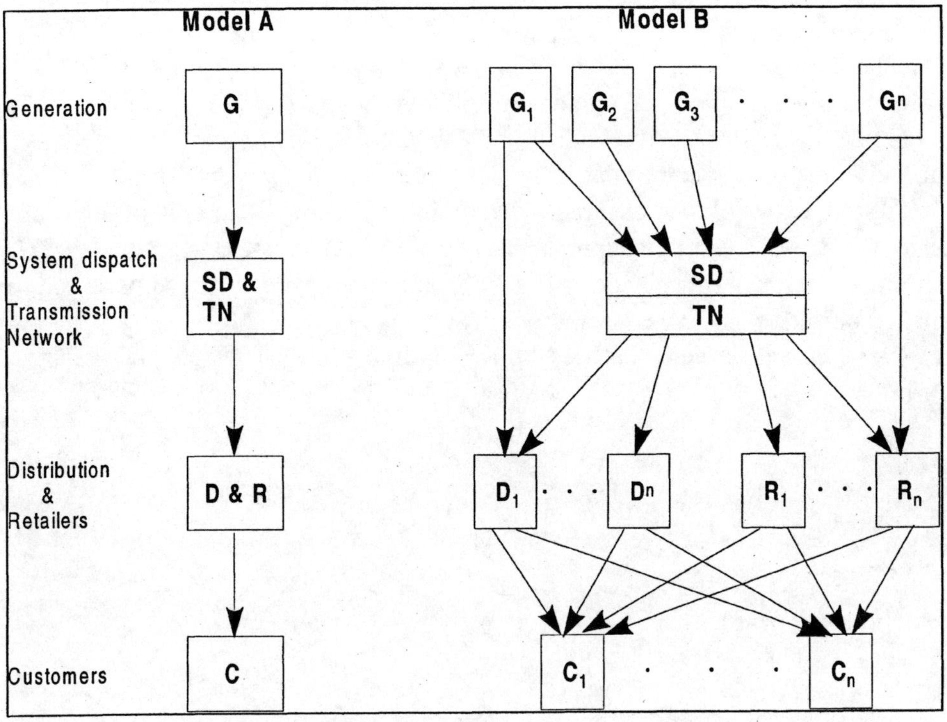


Fig.1 From the monopoly to competitive electricity market

### 3.2 Competition in the Wholesale Market

The reform of Victoria's electricity industry began with the division of the vertically integrated monopoly of the SECV into transmission, generation and distribution entities, as illustrated in Fig. 2.

Victoria was the first state that had introduced reforms leading to the open electricity market. The first stage of the structural reform began in October 1993, when generation, transmission and distribution functions were split and became the responsibility of the three new established companies: Generation Victoria, National Electricity and Electricity Service Victoria [2], [3].

The second stage of the structural reform commenced in October 1994, when five distribution companies, that is City Power, United Energy, Solaris Power, Powercor Australia and Eastern Energy were established. Two new companies are responsible for transmission electrical energy, namely PowerNet Victoria, which owns, maintain and manages High Voltage (HV) networks and trading business, and Victorian Power Exchange (VPX), which is responsible for the operation of the wholesale electricity market. Generation Victoria was broken up to form five independent, corporalised generation businesses consisting of three coal, one gas and one hydro power station groups [2], [3].

All the five distribution businesses were sold in 1995 - 1996 and the remaining generation businesses have been gradually sold during 1996 - 1998. PowerNet Victoria was sold in October 1997. Privatization will lead to the electricity supply market entirely controlled by the private companies being in competition on the open electricity market.

A full-fledged wholesale electricity market introduced in Victoria allows customers to select their own electricity supplier that will offer the most competitive prices and conditions of electricity supply [4]. The companies were first corporatised and significant efficiency gains were achieved.

Figure 2 shows a new, detailed structure of the electricity supply industry in Victoria.

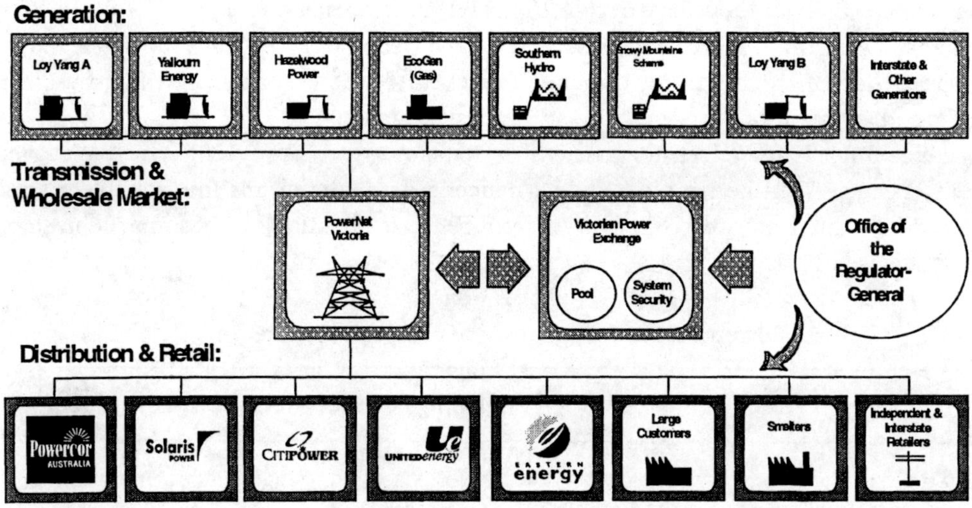


Figure 3. Victorian electricity supply industry structure [1]

Each of the five distribution/retail businesses operates as a 'ring fenced' distribution monopoly and a competitive energy retailer. At corporatisation the distribution assets were valued and the balance sheets then restructured.

The intention was to create viable businesses with the critical mass to maintain prices and service standards of benefit to consumers and attractive to investors because of their size and commercial structure. There are now fourteen retailers operating in Victoria, including the original five, with more expected to join as the NEM increases operation.

A unique feature of reform was the separation of the grid from the wholesale market, system security, and dispatch. Two statutory corporations, PowerNet Victoria (PNV) and Victorian Power Exchange (VPX), were established to operate the transmission system and to operate the wholesale electricity market. PNV owns and maintains the high voltage transmission network. PNV's core function is to ensure the network is maintained and performs to the required standard.

VPX is an independent, non-profit organisation which administers system security and dispatch on behalf of the industry. The prime mandate of VPX is to minimise the cost of electricity to Victoria by matching supply and demand under commercially competitive conditions, while maintaining adequate security and integrity of supply. These key requirements are overseen by the Office of the Regulator-General (ORG).

### 3.3 Competition in the Retail Market

The competition in the retail market has been introduced gradually, with the aim to achieve a fully-fledged market in January 2001. Customers gradually become contestable, that is, they have rights to select an electricity retailer. Currently, non-franchise customers pay for electricity due to the maximum electricity tariffs regulated by the Office of the Regulator-General. These tariffs represent the maximum uniform tariffs and are reviewed every year.

Since July 1998, all customers, which consume more than 160kWh of electrical energy per year will be legible to select their retailer. This means that relatively small businesses with electricity bills around $1,500 per month will become contestable customers, see Table 1.

Table 1. Introduction of a competitive electricity market in Victoria.

| Customers power demand/energy of more than | Selection of Electricity suppliers since |
|---|---|
| 5 MW | December 1994 |
| 1 MW | July 1995 |
| 750 kWh/year and less than 1 MW | July 1996 |
| 160 kWh/year | July 1988 |
| All remaining customers | Since 2000 |

## 4 Privatization of the Electricity Supply Industry in Victoria

The electricity supply industry was gradually sold in 1995 ÷ 1997. The total income from the sale is over 21 billion dollars. Gas power stations are to be privatised soon. Table 2 shows the privatization process.

Table 2. Cash sale proceeds of the Victorian electricity supply industry

| Sector | Name of company | Cash Proceeds [$m] | Date of Sale |
|---|---|---|---|
| Generation | Yallourn Energy | 2,426 | 21 May 96 |
| Generation | Hazelwood | 2,357 | 4 Aug 96 |
| Generation | Loy Yang A | 84 | 1 Apr 97 |
| Generation | Loy Yang B | 4,746 | 22 Apr 97 |
| Generation | Southern Hydro | 409 | Nov 97 |
| Distribution | United Energy | 1,553 | 7 Aug 95 |
| Distribution | Solaris | 950 | 30 Oct 95 |
| Distribution | Eastern Energy | 2,080 | 5 Nov 95 |
| Distribution | Powercor | 2,150 | 16 Nov 95 |
| Distribution | CitiPower | 1,575 | 14 Dec 95 |
| Transmission | Power Net Victoria | 2,716 | 12 Oct 97 |
| ***Total*** | | ***21,046*** | |

## 5 BENEFITS FROM REFORMS

Low-cost energy and efficient infrastructure are two prime requirements for the development of the economy. Victoria's reforms are helping to achieve those goals whilst improving choice and service for consumers, and at the same time reducing the State's public sector debt. In the long term, consumers will benefit from increased pressure on businesses within the NEM to retain customers.

> Proceeds from the sale of the electricity businesses have contributed substantially to reducing public sector debt by more than half in two years. Victoria's ratio of net public sector debt to Gross State Product has fallen from its 1992 peak of 31.3% to 10.6% (estimated at 30 June 1997), the lowest ratio in the State since the 1970s [1] – Fig. 3.

The reform to the electricity supply industry allowed the electricity price reduction in all sectors of energy users. From November 1992 to May 1997 the typical Victorian household which uses 5,500kWh/year benefited from 9.2% real reduction in the cost of electricity. Large industrial and business customers able to choose a retailer achieved substantial savings on their electricity bills. A survey of more than 300 Victorian business by the Australian Chamber of Manufactures showed that 80% of these customers were saving up to 39% and achieving an average 10% reduction in energy costs [1].

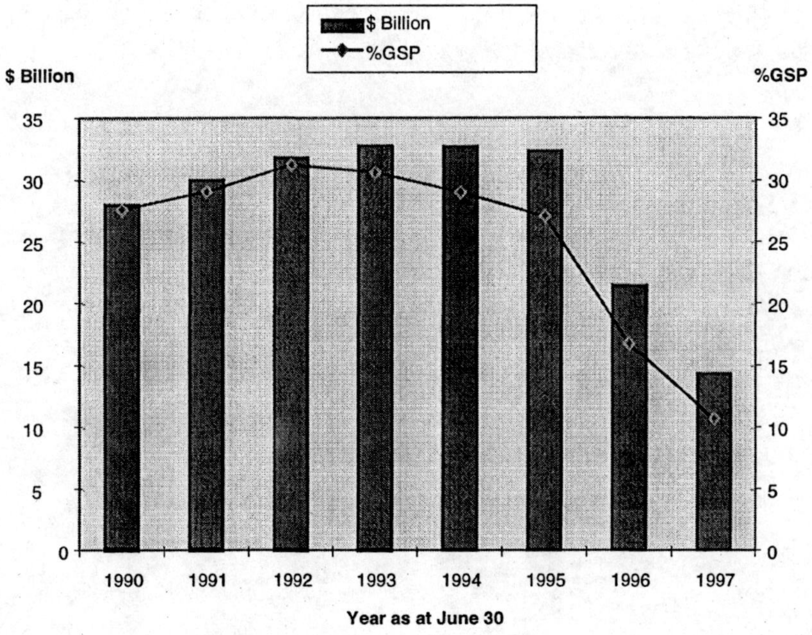


*Source: Victorian Budget Papers 1997*

Fig. 3. Victorian Public Sector debt

Not only lower electricity prices are results of the reform. The Office of the Regulator-General indicates in its report that distribution businesses improved the reliability of supply from 510 minutes off supply per annum under the former monopoly in 1989/90to 218 minutes in 1996 [4].

The Victorian customers noticed more customer orientated approach introduced by competing distributors and retailers. Customer Charters were sent to customers by all power companies. They summarise customer and retailer rights and obligations under the Supply and Sale Code and other key codes and rules.

The Electricity Industry Ombudsman was created as a part of the reformed electricity industry to provide an additional way of customer protection offering resolution of individual complaints and the power to make binding decisions on the electricity supply companies [1].

## 6 Towards the National Electricity Market

The first stage of the NEM commenced on 4 May 1997 allowing trade across borders in integrated markets. The initial participants are Victoria, NSW and the Australian Capital Territory (ACT) with South Australia (SA) participating through the Victorian market. When the full national market commences SA will participate in its own right with Queensland and Tasmania to join later. Consumers are reaping the benefits of this highly competitive marketplace, using the power of choice to obtain better deals on price and service which is expected to continue as all consumers become increasingly able to choose their supplier by the end of the year 2000.

## References

1. *Victoria's Electricity Supply Industry, Towards 2000*, June 1997. Energy Project Division, The State Government of Victoria.
2. *Choice of Electricity Retailer.* Office of the Regulator General, Victoria 1996.
3. VicPool Information Kit. *Victorian Power Exchange*, Victoria 1996.
4. *Electricity Performance Report.* Office of the Regulator-General, June 1997.
5. Hunt, S. and Shuttleworth ,G., "*Competition and Choice in Electricity*," John Wiley & Sons, 1996.

****** 2 ******

# STRUCTURE AND OPERATION OF ELECTRICITY MARKETS


**WLADYSLAW MIELCZARSKI AND GRAZYNA MICHALIK**
Monash University, Department of Electrical and Computer Systems Engineering
Melbourne, Australia, Email: Wlad@basil.eng.monash.edu.au



**Summary:** This paper presents the introduction of the competitive electricity market in Australia. The electricity supply industry in Victoria, the first state which introduced open market is discussed in details. The paper shows a new structure of the electricity supply industry and mutual relations between market participants


## 1 INTRODUCTION

The Australian Electricity Supply Industry (AESI) is one of the largest economic activities in Australia. This industry ranks thirds in terms of its contribution to GDP (2.2% in 1995/95) with $12.3 billion from electricity sales. AESI employs approximately 42000 persons and controls capital assets worth $43 billion. Till 1993, AESI was organised, as the government owned monopolies, which controlled the production, transmission and distribution of electricity in particular states, as seen in Fig. 1. The electricity production reached 160,000 GWh in 1995 with installed capacity of 38,000 GW, as shown in Table 1 [1].

Currently, power supply systems in three states: South Australia (SA), Victoria and New South Wales (NSW) are connected together creating the National Grid. A new AC link between NSW and Queensland (QLD) will be ready in 2001. A DC connection between Victoria and Tasmania is discussed. When all links are constructed, the National Electricity Market will cover five states and Australian Capital Territory, a small enclave with capital Canberra. The Western Australia (WA) and Northern

Territory (NT) will not join the National Market because of large distances from Eastern States.

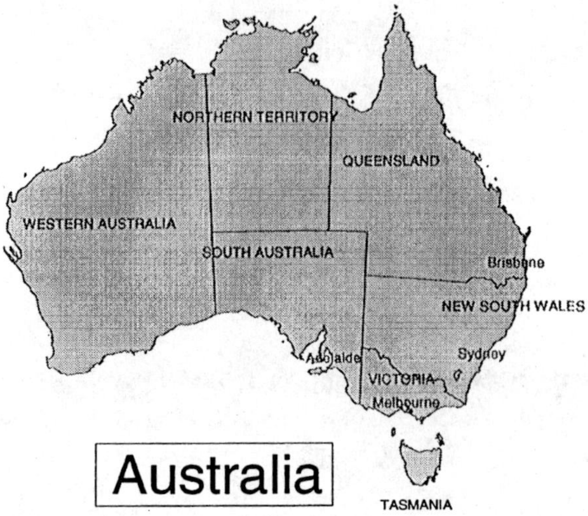


Fig. 1 Australian States and Territories

| *State or Territory* | **Installed Capacity** MW | **Electricity Generation** GWh |
|---|---|---|
| NSW | 12177 | 54339 |
| VIC | 7,154 | 38,174 |
| | | |
| QLD | 6,896 | 31,670 |
| SA | 2,246 | 8,157 |
| WA | 2,957 | 12,177 |
| TAS | 2,509 | 8,698 |
| ACT | - | - |
| NT | 417 | 1,317 |
| SMHEA | 3,756 | 5,583 |
| ***Total*** | ***38,112*** | ***160,115*** |

Table 1 Electricity Generation in Australia, 1995 [1]

State Code used in Table 1:

| | |
|---|---|
| NSW= New South Wales | WA= Western Australia |
| VIC= Victoria | TAS = Tasmania |
| QLD= Queensland | ACT= Australian Capital Territory |
| SA= South Australia | NT= Northern Territory |
| SMHEA= Snowy Mountain Hydro Electric Scheme | |

## 2 Reform to the Electricity Supply Industry

In Australia, electricity supply had been traditionally organized as the regional monopolies in each state and territory, controlled by the Local State Government. The large distances between states and the concentration of population and production around few large centers caused that the electricity supply industry had been operated separately for many years [1].

Since March 1990, four state electricity supply systems (Victoria, NSW, South Australia and ACT) have been interconnected allowing the interstate power transfer. This interconnected system has become the bases for planning the National Electricity Market. In Australia, primarily responsibility for the Electricity Supply Industry rests with the State and Territory Governments; thus, the introduction of the National Electricity Market requires the close coordination of legislation and organization activities of the State Governments. Moreover, the preferences given by the State Governments will vary electricity market in the interconnected states to some degree [1], [2]. Limited interstate power exchange capacities will cause that in the future the national electricity market will face difficulties in exchanging a large portion of electrical energy.

The Special Premiers' Conference of November 1990 began the process leading to the National Electricity Market (NEM) when it was commissioned a study of the feasibility of a national electricity grid. The National Grid Management Council (NGMC) was established six months later and the Council's first draft protocol outlining the principles and objectives of the national market was published by Christmas 1991 [1], [5].

## 3 Electricity Supply Industry Reforms in Victoria

Victoria was the first state that had introduced reforms leading to the open electricity market. The first stage of the structural reform began in October 1993, when generation, transmission and distribution functions were split and became the responsibility of the three new established companies: Generation Victoria, National Electricity and Electricity Service Victoria [3], [6].

The second stage of the structural reform commenced in October 1994, when five distribution companies, that is City Power, United Energy, Solaris Power, Powercor Australia and Eastern Energy were established. Two new companies are responsible for transmission of electrical energy, namely PowerNet Victoria, which owns, maintain and manages High Voltage (HV) networks and trading business, and Victoria Power Exchange (VPX), which is responsible for the operation of the wholesale electricity market. Generation Victoria was broken up to form five independent, corporalised generation businesses consisting of three coal, one gas and one Hydro power station groups [3], [7].

All the five distribution businesses were sold in 1995-1996 and the remaining generation businesses have been gradually sold during 1996-1998. PowerNet was sold in 1997. Privatization will lead to the electricity supply market entirely controlled by the private companies being in competition on the open electricity market.

A full-fledged wholesale electricity market introduced in Victoria allows customers to select their own electricity supplier that will offer the most competitive prices and conditions of electricity supply [4]. Table 2 shows the stages of the reform leading to the selection of electricity suppliers.

Table 2 Introduction of a competitive electricity market in Victoria

| **Customers power demand/energy of more than** | **Selection of Electricity suppliers since** |
|---|---|
| 5 MW | December 1994 |
| 1 MW | July 1995 |
| 750 kWh/year and less than 1 MW | July 1996 |
| 160 kWh/year | July 1988 |
| All remaining customers | Since 2000 |

The State Government still controls energy prices for households, farms, and small sized business by the electricity tariffs. The Victorian Government has established the institution, The Office of the Regulator-General (ORG), that acts as an independent regulator in order to monitor, maintain, and integrate the electrical energy supply. The ORG is also responsible for the maximum values of network charges and the maximum tariffs.

## 4 New Structure of Electricity Supply Industry in Victoria

In 1993, the Victorian Government introduced a major restructuring of the electricity industry in three distinct stages [6] leading to a new structure of the electrical supply industry (see Fig 2).

The most important element of the reform is the separation of the distribution from generation and the buying and selling (retailing) of electricity. All generation and distribution companies use the same transmission and distribution networks. The cost of using a network is the same for the owner as for any other supplier. Customers are able to negotiate competitive prices for electricity and purchase electrical energy from suppliers that offer the most attractive prices and conditions.

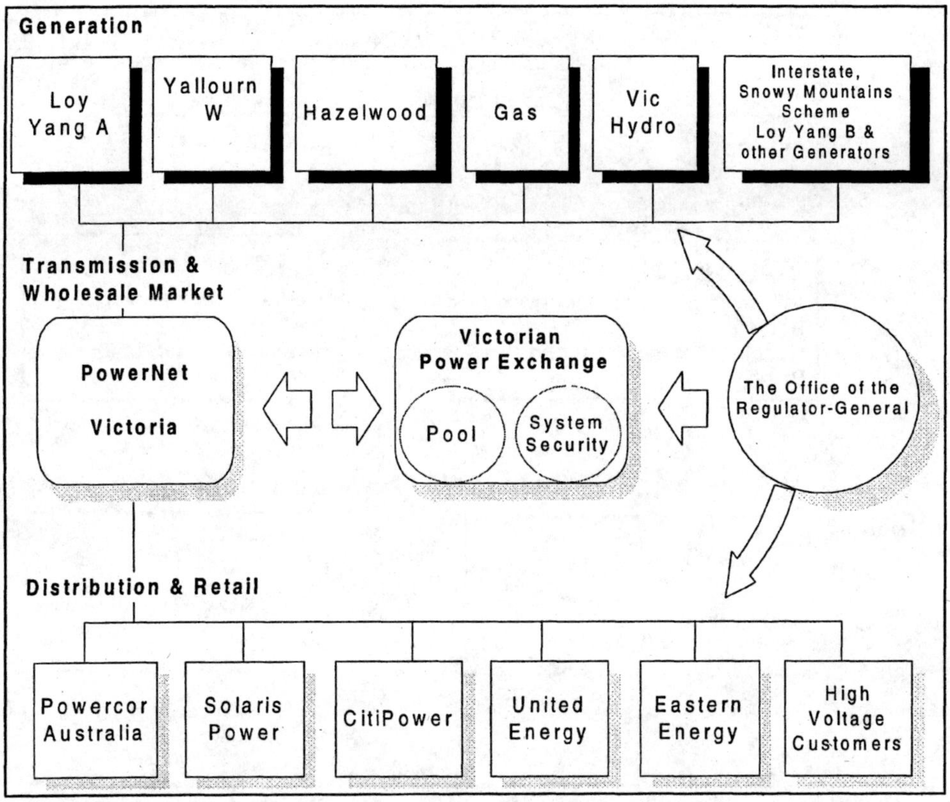


Fig. 2 New structure of the electricity supply industry in Victoria

A competitive electricity market allows everybody approved by the Office of the Regulator-General (ORG) to buy and sell electricity in the spot market. The ORG oversees the operation of the pool to ensure its competitive operation. A structure of a competitive electricity market in Victoria is shown in Fig. 3 [6].

Competition, introduced in the electricity supply industry, operates through the following mechanism:

1. ***Wholesale Market*** created by the independent multiple buyers (the five Distribution companies, retailers, large consumers, and interstate purchases) and multiple sellers.
2. ***Retail competition*** in which the retail price for deregulated customers is contestableand these customers may select suppliers. The number of customers with choice at the retail level increases towards the year 2000 when all customers will be allowed to select a retailer.

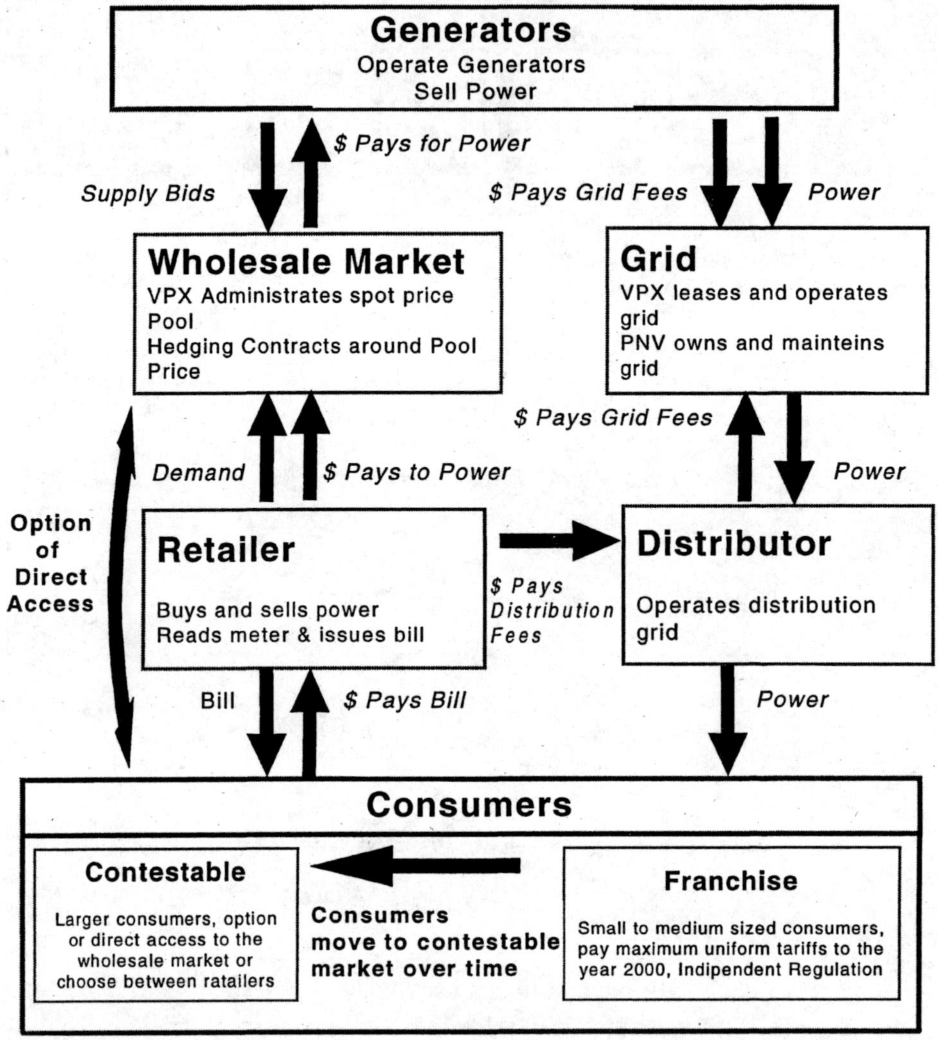


Fig. 3 Structure of the competitive electricity market in Victoria

3. ***Competition in input markets*** allowing competitive tendering for other services including maintenance and construction.

The market has two basic elements:

***The Pool*** which allows to set prices for the next day. Each day generators bid into the Pool the price of electricity for half-hour in the following day. The next day, generators are dispatched to meet the market demand from buyers in the order of their bids. The lowest price is dispatched first. The price of the highest priced generators, which is actually dispatched in a given period, set the Pool price for that period. This price is received by all generators dispatched in that period.

***Hedging contracts*** which are financial instruments between the wholesale Buyers and Sellers. These contracts allow managing financial risk inherited in varying Pool prices.

## 5 Bids and Offers

The spot price of electricity in the market is established by equating demand with supply. Both producers and retailers enter their bids and offers for each day, and each trading interval. In the 24 hours before 10:00 am every workday and in the 2 hours before 1:00 pm every workday, each participant may submit a new bid covering the period up to 7 days. The generators can change their bids after they have been submitted. However, they have to decide whether to commit specific units 41 hours before the trading interval. Two days before the trading day VPX uses the bids and offers to estimate the price for that trading day. After that time participants cannot change their commitment. However, up to 90 minutes before each trading interval a generator can change the band prices of its bids.

Both retailers and electricity users can offer the reduction of the load to the pool on the same basis as generators offer energy production. The offer of load reduction is called Demand Management unit.

Operation of the electricity pool involves a number of steps in an overall process:

- Generators bid prices for their generation units to supply the Market for the scheduled period (currently seven days ahead). These bids take into consideration the capabilities of those generators (eg. minimum load, how long it takes to start generating etc)
- Demand for electricity is estimated for the upcoming period, taking into account such features as weather, special events, seasonal characteristics, holidays, etc
- Generators are scheduled in "merit order", the cheapest of the bidders being used to supply the demand for electricity
- Checks are made to ensure system security requirements are met, and schedules may need to be altered to satisfy system security criteria
- Generation is then dispatched to meet the electricity demand as it varies throughout the day

### 5.1 Band Levels

In their bids, the generators provide supply levels of each unit for each band. Table 3 gives an example of band levels for one generating unit.

Table 3 An Example of Band Levels for Generating Unit

| Time | Unit | Band1 | Band2 | Band3 | Band4 | Band5 | Band6 | Band7 | Band8 | Band9 | Available capacity [MW] |
|---|---|---|---|---|---|---|---|---|---|---|---|
| **0:00** | AAA | 270 | 60 | 100 | 70 | 0 | 30 | 30 | 30 | 30 | 680 |
| **0:30** | AAA | 270 | 60 | 100 | 50 | 0 | 30 | 30 | 30 | 30 | 680 |
| **1:00** | AAA | 270 | 60 | 100 | 0 | 0 | 30 | 30 | 30 | 30 | 680 |
| **1:30** | AAA | 270 | 60 | 70 | 0 | 0 | 30 | 30 | 30 | 30 | 680 |
| **2:00** | AAA | 270 | 60 | 40 | 0 | 0 | 30 | 30 | 30 | 30 | 680 |
| **2:30** | AAA | 270 | 60 | 30 | 0 | 0 | 30 | 30 | 30 | 30 | 680 |
| **3:00** | AAA | 270 | 60 | 10 | 0 | 0 | 30 | 30 | 30 | 30 | 680 |
| **3:30** | AAA | 270 | 60 | 10 | 0 | 0 | 30 | 30 | 30 | 30 | 680 |
| **4:00** | AAA | 270 | 60 | 10 | 0 | 0 | 30 | 0 | 30 | 30 | 680 |
| **4:30** | AAA | 270 | 60 | 20 | 0 | 0 | 30 | 0 | 30 | 30 | 680 |
| **5:00** | AAA | 270 | 60 | 70 | 0 | 0 | 30 | 0 | 30 | 30 | 680 |
| **5:30** | AAA | 270 | 60 | 90 | 0 | 0 | 30 | 0 | 30 | 30 | 680 |
| **6:00** | AAA | 270 | 60 | 100 | 40 | 0 | 30 | 0 | 30 | 30 | 680 |
| **6:30** | AAA | 270 | 60 | 100 | 40 | 0 | 30 | 0 | 30 | 30 | 680 |
| **7:00** | AAA | 270 | 60 | 100 | 50 | 50 | 30 | 0 | 30 | 30 | 680 |
| **7:30** | AAA | 270 | 60 | 100 | 50 | 70 | 30 | 0 | 30 | 20 | 680 |
| **8:00** | AAA | 270 | 60 | 100 | 50 | 50 | 30 | 0 | 30 | 30 | 680 |
| **8:30** | AAA | 270 | 60 | 100 | 50 | 60 | 30 | 0 | 30 | 30 | 680 |
| **9:00** | AAA | 270 | 60 | 100 | 50 | 100 | 30 | 0 | 30 | 20 | 680 |
| **9:30** | AAA | 270 | 60 | 100 | 50 | 90 | 30 | 0 | 30 | 20 | 680 |
| **10:00** | AAA | 270 | 60 | 100 | 50 | 60 | 30 | 0 | 30 | 30 | 680 |
| **10:30** | AAA | 270 | 60 | 100 | 50 | 50 | 30 | 0 | 30 | 30 | 680 |
| **11:00** | AAA | 270 | 60 | 100 | 50 | 30 | 30 | 0 | 30 | 30 | 680 |
| **11:30** | AAA | 270 | 60 | 100 | 50 | 50 | 30 | 0 | 30 | 30 | 680 |
| **12:00** | AAA | 270 | 60 | 100 | 50 | 80 | 30 | 0 | 30 | 30 | 680 |
| **12:30** | AAA | 270 | 60 | 100 | 50 | 80 | 30 | 0 | 30 | 30 | 680 |
| **13:00** | AAA | 270 | 60 | 100 | 50 | 80 | 30 | 0 | 30 | 30 | 680 |
| **13:30** | AAA | 270 | 60 | 100 | 50 | 80 | 30 | 0 | 30 | 30 | 680 |
| **14:00** | AAA | 270 | 60 | 100 | 50 | 80 | 30 | 0 | 30 | 30 | 680 |
| **14:30** | AAA | 270 | 60 | 100 | 50 | 80 | 30 | 0 | 30 | 30 | 680 |
| **15:00** | AAA | 270 | 60 | 100 | 50 | 70 | 30 | 0 | 30 | 30 | 680 |

Secondly, the generators submit prices for each of the bands entered. Along with the prices, the generators also submit bid elbows - if a specific MW level is reached then the generators are only willing to supply more electricity at a higher price.

## 5.2 Bidding Behaviour of Power Stations

Generators provide information to the pool managers on their commitment to the pool in three main parameters:

- energy prices
- power offered in particular trading intervals
- plant/units availability

Energy prices are determined in 10 price bands and are submitted to the pool manager a day before the trading day. There is only one limitation when determining price bands, namely, it is required that every consecutive price band has a higher price. However, this requirement seems to be unjustified. In practice, a generator is producing some amount of electricity for a price 'x'", while the excess of energy could have been offered for a price 'y', where x > y. Unfortunately, the market regulations do not allow the reduction of prices along the price band. Examples of a price bands are shown in Figures 4 and 5. An example of a thermal power station offers is shown in Fig. 4. Four units are offered for similar prices. Due to regulations, the prices bidded have to increase along the bands. In some cases, this requirement is achieved by the price increase in one-cent steps when a power station wants to offer energy for the same price.

The bid of a gas power station presented in Fig. 5 is typical for pool units. One gas turbine is offered for a relatively low price, while the others are usually offered when a spot price is close to the Value of the Lost Load (VoLL = A$5,000) in order to maximise the profit. Because high cost of fuel, gas power stations are not able to compete with thermal units when supplying basic load for low spot prices.

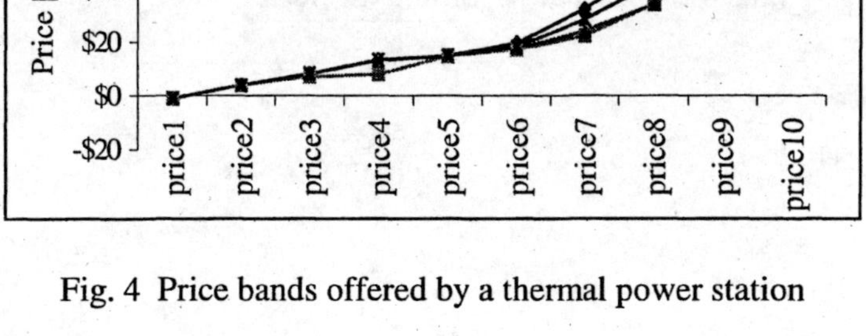


Fig. 4 Price bands offered by a thermal power station

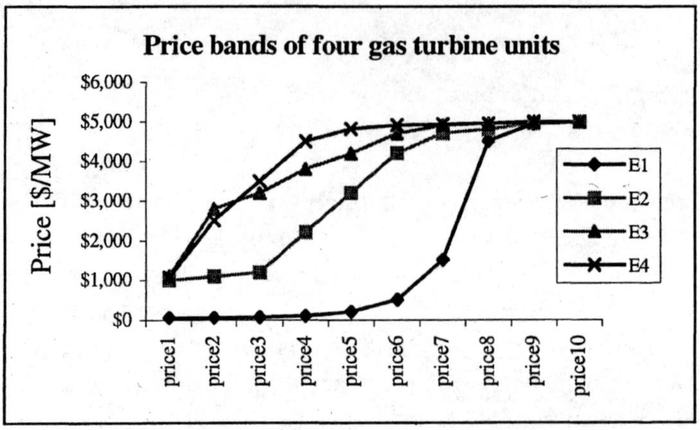


Fig. 5 Price bands offered by a gas power station

Figure 6 shows the bids of a thermal unit. The first portion of the energy (300MWh) is offered as a self committed unit. Therefore, due to regulations, its price must be negative and in this case equals to - $1/MWh. The second portion (about 150 MWh) is offered for $4.04/MWh. The next price band is $8.34/MWh and about 50MWh is offered. No electrical energy is offered for $13.34/MWh. The next price where energy is offered is $15/MWh. Additionally, this power station offers some amount of energy during off-peak hours for $49/MWh. The average spot price expected for this day is around $12/MWh.

These bids are typical for thermal units that are fully covered by vesting and bilateral contracts. The bids of power station considered are very low in order to be scheduled and generate due to the contract signed. If this power station was not generated and spot prices increased above the contract strike prices, the power station would have to compensate the difference to its partners. This would result in losses of revenue.

Higher cost of fuel in Region 2 allows thermal power stations from Region 1 to export electrical energy for higher prices during off-peak hours. The power station presented in Fig. 6 wants to take this opportunity, offering energy for $49/MWh during off-peak trading periods.

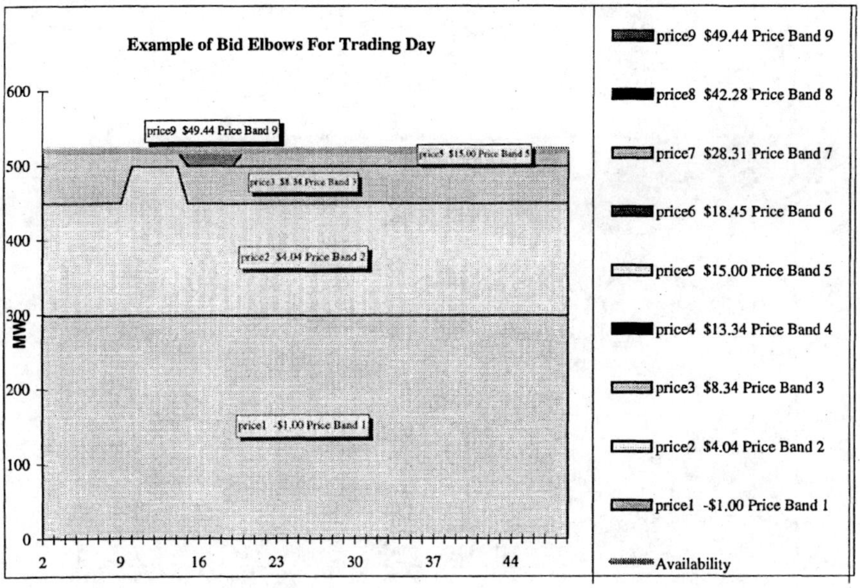


Fig. 6 Bids of a thermal power station

Customers submit bids that contain the amount of energy they require for each trading interval and the price they are willing to pay for it.

Bidding behaviour of gas power station is shown in fig. 7. This power station consists of seven gas turbine units but only one has been selected for presentation because the bidding offers of the remaining six units are identical. The cost of energy generation for gas turbine units is at least 4 ÷ 5 times larger than the average spot price level. This power station does not have long-term hedge contracts, since practically it would be impossible for this type of power station to offer competitive prices. The gas power station plays in the pool offering energy for very high price, that is, $4,962.72/mwh. This price is just below the maximum price level, which is determined by the value of lost load (voll = $5000/mwh). It seems that this strategy is the only possible method to survive on the competitive market for energy-peak units with high cost of fuel when spot prices are generally very low.

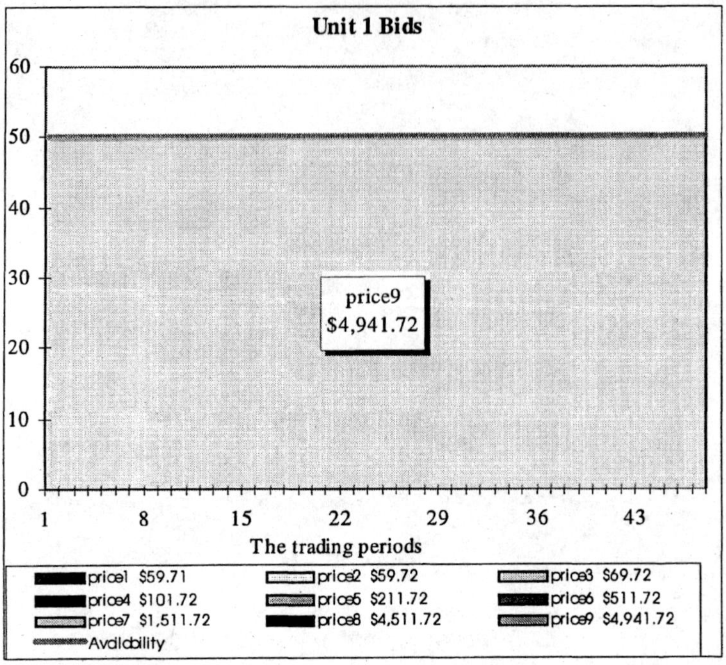


Fig. 7 Bidding behaviour of a gas turbine power station in Region 1

Power stations can offer in energy pool commitment of their real, physical units, they can combine a few units in one or can create 'artificial' units, called notational units, in which energy offered is generated by various combinations of real units' commitments. The bidding strategies of generators with notational units require a lot of experience, especially when there is a low level of vesting and hedge contracts, forcing the power station to play more sophisticated game on the energy market.

## 6. Towards the National Electricity Market

The success of the Victorian electricity market facilitated the development of the National Electricity Market (NEC). The first stage of the National Electricity Market started in May 1997 and is known as NEM1. The second stage is to be started in April 1998. Gradual development of the NEC and connection of the Queensland power system, with a possible DC link to Tasmania will create in 3÷-5 years one of the largest open electricity market.

### Market Overview

The National Electricity Market Management Company (NEMMCO) will operates a spot market (the Pool) for the wholesale trading of electricity. The wholesale electricity market operates in accordance with the rules specified within the National Electricity Code introduced by National Electricity Code Administrator (NECA). Sales and purchases of electricity in this market are made between participating Generators and Suppliers according to a set of rules that govern the market's operation and the calculation of payments. A computerised system (the Settlement System) is applied to calculate prices and to process metered, operational and other data required to determine the payments.

NEMMCO will also facilitate a Short Term Forward Market that enables participants to manage short-term changes in expected generation and demand and adjust their electricity contract exposure accordingly. In addition, NEMMCO will provide the facility for trading Inter-Regional Hedging Contracts. The main organisations involved in the Market are as shown in the Fig. 8 [2].

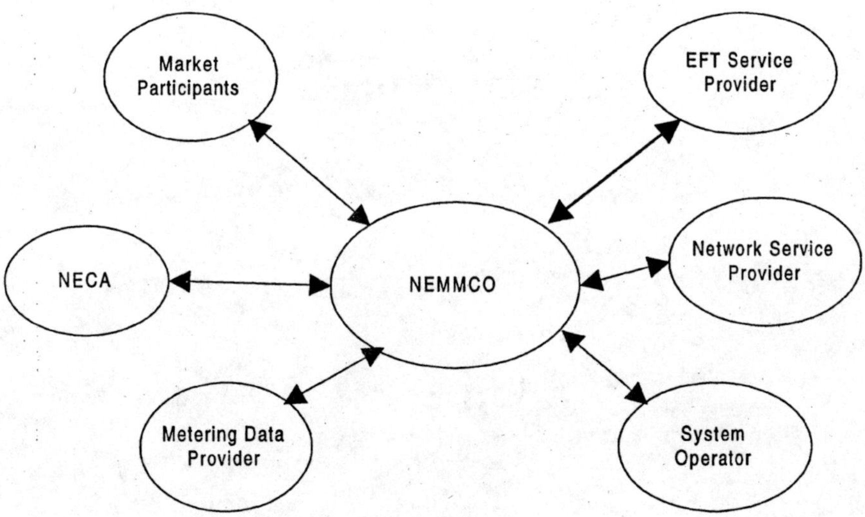


Fig. 8 Main organisations involved in the National Electricity Market

Current structure of the National Electricity Market is shown in Fig. 9.

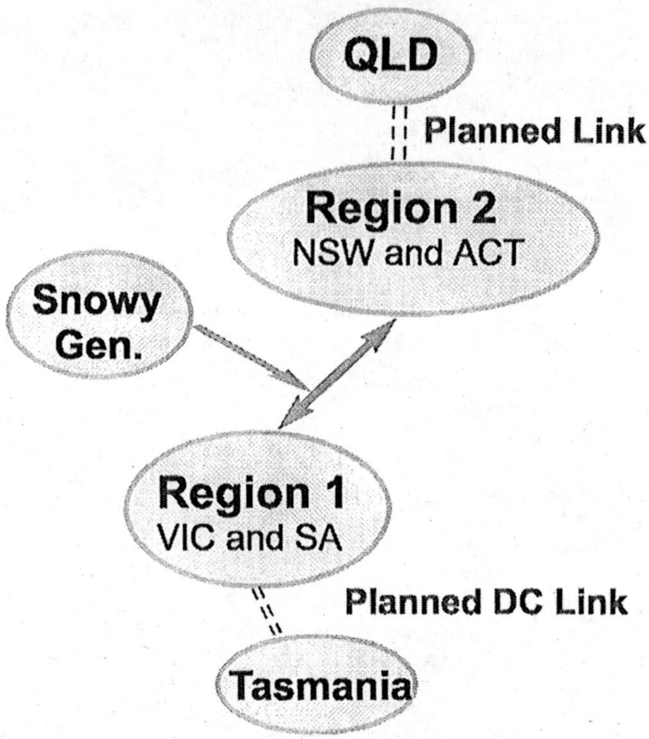


Fig. 9 Structure of the National Electricity Market

Notation:

VIC = Victoria
SA = South Australia
NSW = New South Wales
ACT = Australian Capital Territory
QLD = Queensland
Snowy Gen. = Snowy Mountains Generation Scheme

The NEC is presented in details in the article "The National Electricity Market, its Structure and Development," published in this issue.

## REFERENCES


1 *Electricity 1996*, International Benchmarking, The Productivity Commission, The Commonwealth Government of Australia, 1996.

2 *Australia's National Electricity Market – An Introduction*, National Electricity Market Management Company Limited, 1997.

3 The Electricity Supply Industry in Victoria. *A Competitive Future – Electricity*. Department of the Treasury, The Victorian Government, 1995.

4 *Choice of Electricity Retailer*. Office of the Regulator General, Victoria 1996.
5 Empowering the Market. *National Grid Management*, 1995.
6 *Electricity Industry Regulatory Statement*. Office of the Regulator General, Victoria, 1995.
7 *VicPool Information Kit*. Victorian Power Exchange, Victoria 1996.

****** 3 ******

# WHOLESALE ELECTRICITY MARKET

**WLADYSLAW MIELCZARSKI AND GRAZYNA MICHALIK**
Monash University, Department of Electrical and Computer Systems Engineering
Melbourne, Australia, Email: Wlad@basil.eng.monash.edu.au

**Summary** The paper presents the wholesale electricity market in Victoria. There are several categories of participants involved in the wholesale market: the pool manager (Victorian Power Exchange), the transmission company (Power Net Victoria), electricity generators, distribution businesses and electricity retailers and traders. Characteristics of market participants and relations among them are discussed.

## INTRODUCTION

Major restructuring of Victoria's electricity supply industry has been the State Government's initiative. The operation of the market and the power system is facilitated by the organisations:

***Victorian Power Exchange (VPX)** - is* responsible for the operation of the power system and the Wholesale Electricity Market

***PowerNet Victoria (PowerNet)*** - owns and maintains the extra high voltage (transmission) network

***Distribution Wires Businesses*** - own and maintain the distribution network

The open electricity market in Victoria (VicPool) began in July 1994. It was transformed into the National Electricity Market (NEM) in May 1997, when three States - New South Wales, Victoria and South Australia - established the joint electricity market. There are two distinct periods of the electricity market in Victoria:

- from July 1994 to May 1997 (the VicPools)

- from May 1997 the National Electricity Market (NEM1).

Relationships among wholesale market participants are shown in Figure 1.

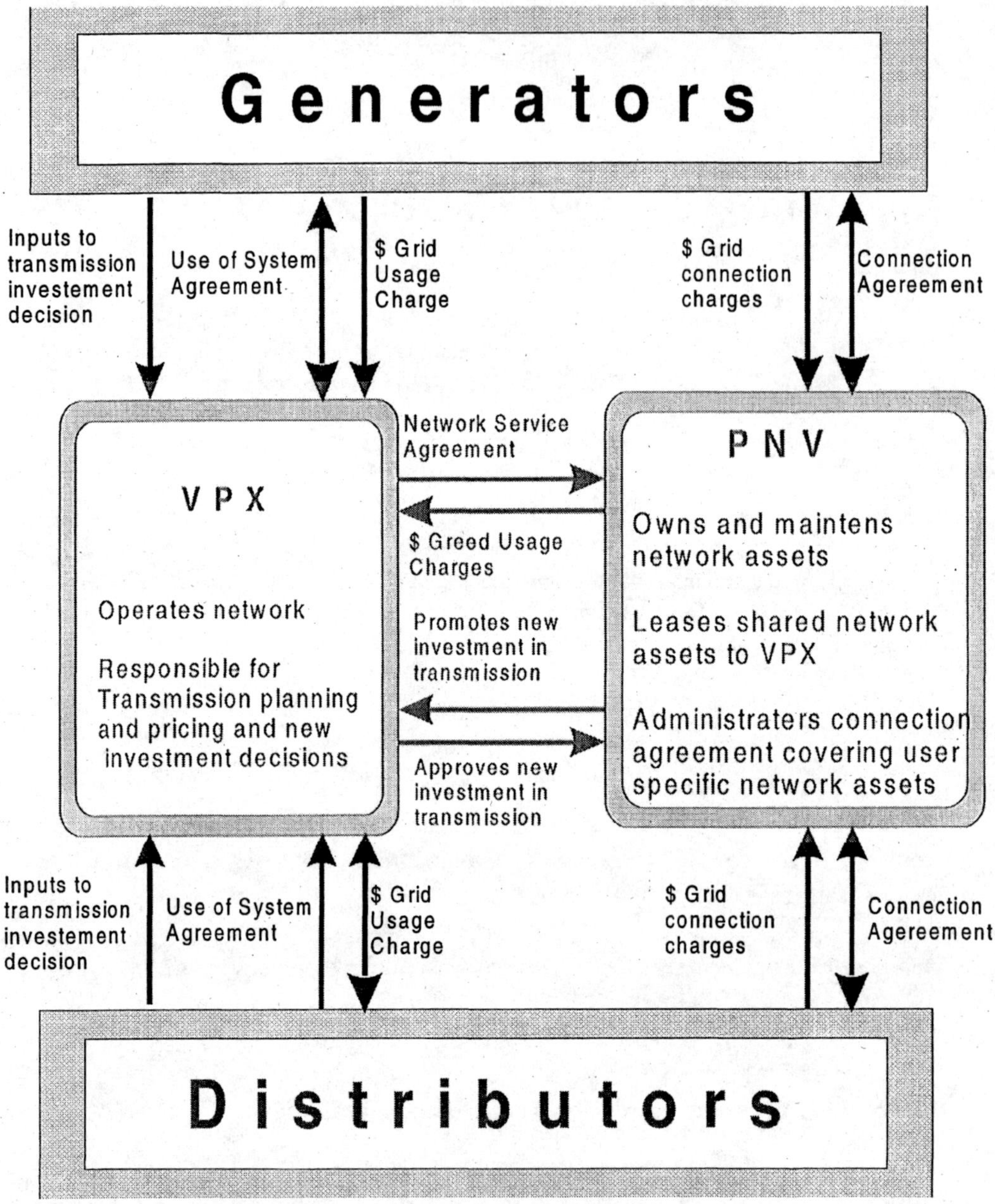


Figure 1. Relationships in the wholesale electricity market in Victoria.

## 2 VICTORIAN POWER EXCHANGE (VPX)

The Victorian reform has separated the generation and retail sectors, and the main grid. Victorian Power Exchange (VPX) has been established to manage the security of the State's power system, manage the Wholesale Electricity Market and plan and direct the development of the Extra High Voltage transmission system. VPX was introduced under the Electricity Industry (Amendment) Act of 1994 and formally came into existence on 3 October, 1994.

VPX is a statutory, independent business. It links all electricity supply industry businesses both physically and financially. Physically, VPX is responsible for the secure operation of the main power system. Its financial responsibilities include managing spot trading in the Wholesale Electricity Market and this includes the very important settlement of market transactions.

VPX's role includes:

- operating and administering the market for spot trading of electricity;
- controlling generation dispatch, and matching available generation to demand;
- controlling the security of the main power system;
- operating the electricity transmission system;
- planning and directing the augmentation of the Extra High Voltage transmission system; and
- providing information to market participants to enable them to operate effectively in the new market.

VPX must maintain the system security a level required by market participants allowing reliable delivery of electrical energy to the community [1].

## 3 POWER NET VICTORIA

Power Net Victoria (PNV) is a statutory corporation created under the *Electricity Industry Act 1993*. PNV is a fully commercial business with an independent Board. PNV owns and maintains the existing network assets. The corporation's core function is to ensure its assets are adequately maintained and plant performance meets the required standards. The company also aims to achieve optimal performance of its assets. The corporation's assets include: transmission lines and substations down to and including 66 kV, static Var sources; and transmission protection systems associated with the network assets [5].

The PowerNet transmission system covers an area of 227,660 square kilometres including 6,740 circuit kilometres of overhead lines and 10 kilometres of underground 220 kV cable connect the system. The system serves the 4.5 million people in Victoria. The overhead lines owned by PNV are as follows:

550 kV - 1,516 km
330 kV - 739 km
275 kV - 156 km
220 kV - 4,177 km
66 kV - 142 km

The Power Net interface with the five distribution companies. The connections are mainly at 66 kV and at 22 kV or 11 kV. The generation interfaces with the PowerNet system at the high voltage terminals of the generator transformers. PowerNet has terminal stations at 43 sites.

Power Net assets include 9,765 MVA of 66 kV and 22 kV load supplying transformation to the distribution companies and 9,480 MVA of transformation between the transmission voltages of 500 kV, 330 kV, 275 kV and 220 kV.

Under the terms of a contract between PNV and VPX called the *Network Service Agreement*:

- PNV bills VPX for use of the shared network assets that are not attributable to individual users, in return for a specified charge;
- PNV is responsible for providing transmission assets of defined capabilities and availability; communications and system monitoring services for system control purposes, and metering services which are installed and maintained to a defined standard; and
- VPX is responsible for using the assets within limits specified by PNV.

The relationship between PNV and VPX can be summarised as follows [5]:

**Network assets and operations.** VPX operates PNV's network assets that are shared and not specifically attributable to individual users, in return for payment of a regulated charge for usage for the assets.
**Connection assets.** PNV connects distributors/large customers and generators to the high voltage grid and recovers a grid connection charge under a *Connection Agreement* for the provision of user specific transmission assets.
**New investment.** VPX analyses the optimal balance between the costs (to market participants) of transmission constraints versus new investment in transmission assets. In this activity, VPX draws on information from generators, distributors, large customers and PNV. VPX provides approval for any network investment.

The relationships between Power Net and other market participants are shown in Figure 2.

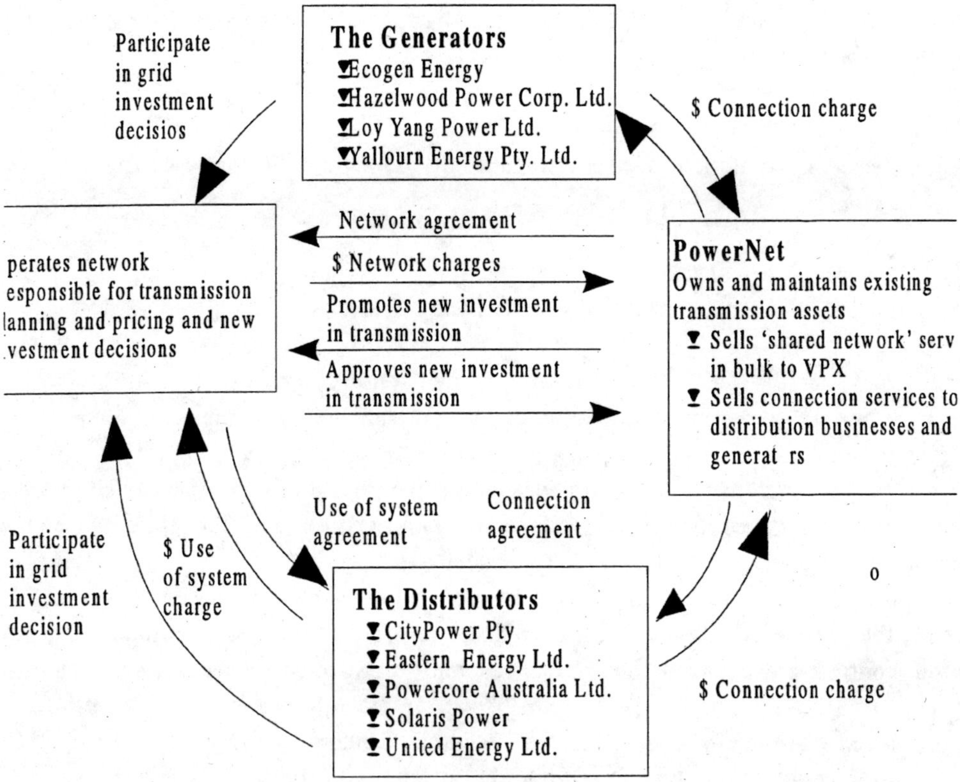


Fig.2 Relation between Power Net Victoria, VPX, generators and distributors [5].

In October 1997, Power Net Victoria was sold to the American company GPU.

**VICPOOL**

VicPool is the name of the competitive market which has been operating in Victoria since 3 July 1994. The market is a spot price market where the wholesale or "pool" price for electricity can vary on a half hourly basis, depending on the electrical power system demand, bid prices and availability of generation and transmission.

VicPool is a transitional market arrangement where the time and resources available to establish facilities have restricted its current implementation. There will be ongoing enhancements to the market as the rules are developed and new systems implemented. Vicpool will be changed into one trading region within the National Electricity Market.

Operation of VicPool involves a number of steps in an overall process:

Generators bid prices for their generation units to supply the Market for the scheduled period. These bids take into consideration the capabilities of those generators (eg. minimum load, how long it takes to start generating etc) Demand for electricity is estimated for the upcoming period, taking into account such features as weather, special events, seasonal characteristics, holidays, etc

Generators are scheduled in "merit order", the cheapest of the bidders being used to supply the demand for electricity
Checks are made to ensure system security requirements are met, and schedules may need to be altered to satisfy system security criteria
Generation is then dispatched to meet the electricity demand as it varies throughout the day

## MANAGING ENERGY MARKET

Participants in the market are the generators and buyers:

***generators*** - supply energy from generating units at power stations
***pool customers*** - are either retail suppliers or non-franchise customers:

Generation businesses submit a bid for each of their generating units, advising how much electricity they are prepared to produce and the price at which they are prepared to produce electrical energy.

Customers are able to bid an amount by which they are prepared to reduce their demand if the pool price rises above a nominated level. This reduction in load must be under the control; however, customers can respond to expected prices without placing bids [2].

VPX is responsible for scheduling the available generation resources to meet the demand and reserve requirements at minimum cost over the schedule period. The schedule is based on the generator bids. The major stages of scheduling are assessing demand and determining the mix of generator and demand side bids to meet the market at lowest cost.

VPX forecasts the customer demand for electricity based on weather forecasts and historical records of demand. Then it selects the lowest price combination of generation and demand side bids to meet the normal customer demand. To ensure that the lowest price combination of generating units is available for operation across the week, a unit commitment analysis is carried out to determine if additional thermal units should be started up or other units shut down.

Closer to the time, an economic dispatch process is used to allocate loading of generating units or instructions to demand side bidders to reduce demand. The economic dispatch starts with the lowest priced generation or demand bids and progressively loads higher priced bids until customer demand is met. [2].

The half hourly pool price is determined "after the even" and is set at the price of the highest bid used in the half hour. This is known as the System Marginal Price (SMP) or Pool Price. All wholesale electricity is traded through the pool at the pool price. All generators are paid the pool price for their energy. Generators may also be paid for technical services and special cases.

**WHOLESALE METERING**

To calculate the Pool price and enable financial settlement for energy traded, it is essential that energy flows at all connection points between market participants and the transmission and distribution networks are accurately measured. Metering facilities must meet the requirements of the Wholesale Metering Code issued by the Office of the Regulator-General.

Physically, over 400 sophisticated, computerised meters were installed before VicPool started operating. These are located in the transmission network's main Terminal Stations (at 500kV to 66kV) at the connection points for each generator, interstate connection and points of supply to the distribution businesses and non-franchise customers (eg. large industrial loads). Meters are also located in the distribution networks' Zone Substations (at 66kV to 11 kV) at connections points between the five Distribution Businesses. "Check" meters have been installed at key locations to enable the validation of measurements and to provide an alternate source of data in case of failure of the Official meter [3].

The following data is collected for each quarter hour for each day, from each

Active and Reactive energy flows (Wh and VArh).

The time, location and direction of energy flows.

Status information on the operation of the meter.

The meters continuously measure and store energy flow data over each 15 minute period and up-load the information to VPX's central database (the Metering Database or MDB) every 24 hours. A meter accuracy of +/- 0.2% and synchronisation of measurements within two seconds of Eastern Standard Time are required by the Wholesale Metering Code.

The meters are computer based measuring and recording devices that use high speed sampling and digital conversion techniques. The compensation of measurement transformers is carried out in software. Sixty two days of data is stored in each meter, to allow time for completion of the Pool Settlement process before the source data is lost. Data and meter operation are monitored by in-built validation processes which produce alarms if a problem exists [3].

Meters at each site are linked by fibre-optic Local Area Networks and a nominated 'master' meter, which collates data from the others, is "polled" by the Metering Database. Access to the metering database is available to all Pool participants via Electronic Mail. The database control allows the participants to have only access to the data relating to their own organisation [3].

## 4.2 SYSTEM CONTROL

The Victorian Transmission Network is interconnected with South Australia, and New South Wales, forming the South East Australian Grid. This interconnected power system has a maximum demand of over 19,000 Megawatts (MW's), and annual energy generated over 100,000 GigaWatthours.

The Victorian Power System has:

- installed capacity participating in the VicPool market of 8,260 MW (including Victoria's share of the Snowy Scheme).
- an expected peak demand of over 7100 MW (1998)
- a network of 70,000 kms of high voltage transmission lines to distribute electricity throughout the State.

VPX's System Control Centre (SCC) controls the Victorian power system. The SCC is responsible for operating VicPool, securely operating the power system, and for monitoring and controlling the system under normal and abnormal conditions. System Control Centre staff direct operations of the Victorian transmission network through VPX's Regional Control Centres, and schedule and dispatch generation to meet the needs of customers [4]

Data required for power system management (eg. generation levels, power line flows, alarm signals, weather details) arrive at System Control Centre from locations around the State (and interstate) via power line carrier, microwave, fibre optic links and radio. Data is updated regularly, some every 2 seconds. This data is used by the System Control Centre Energy Management System (EMS) for a wide range of functions. The communication links are also used for control and monitoring of a range of equipment.

## POOL CHARGES

Victorian Power Exchanges charges pool participants a pool fee to cover costs of the market management and metering. The main components of the pool fee for a financial year 1997-98 include:

| | |
|---|---|
| Participant Membership Fee | $30 per day |
| Wholesale Meter Fee | $7.00 per meter per day |
| Pool Fee (pool customer settlement energy) | $0.247/MWh |
| Ancillary Services Fee (pool customer settlement energy | 0.522/MWh |
| Hot Wire Business charge (for obtaining and processing metering data) | $522/meter/year |
| Smelter Reduction Amount Rate (1) (franchise energy) | $2.00/MWh |
| Smelter Reduction Amount Rate (2) (non franchise energy) | 2.09/MWh |

Since the 1980s the State Electricity Commission of Victoria (SECV) has had long-term agreements with Alcoa to supply electricity to the Portland and Point Henry Smelters, which consume greater than 720 MW of base load electricity supplied under these contracts. These contracts are managed by the Smelter Trader. [5]. To reduce the Government losses under these contracts a special levy was introduced on the wholesale market participants. This levy is known as Smelter Reduction Amount Rate.

## GENERATORS

On 31 January 1995 Generation Victoria, was disaggregated into five independent and commercially viable generating businesses, with the SECV retaining the Government's interests in Loy Yang B and the Snowy Mountains hydro-electric scheme. The structure of the Victorian generation is shown in Figure 3 [5].

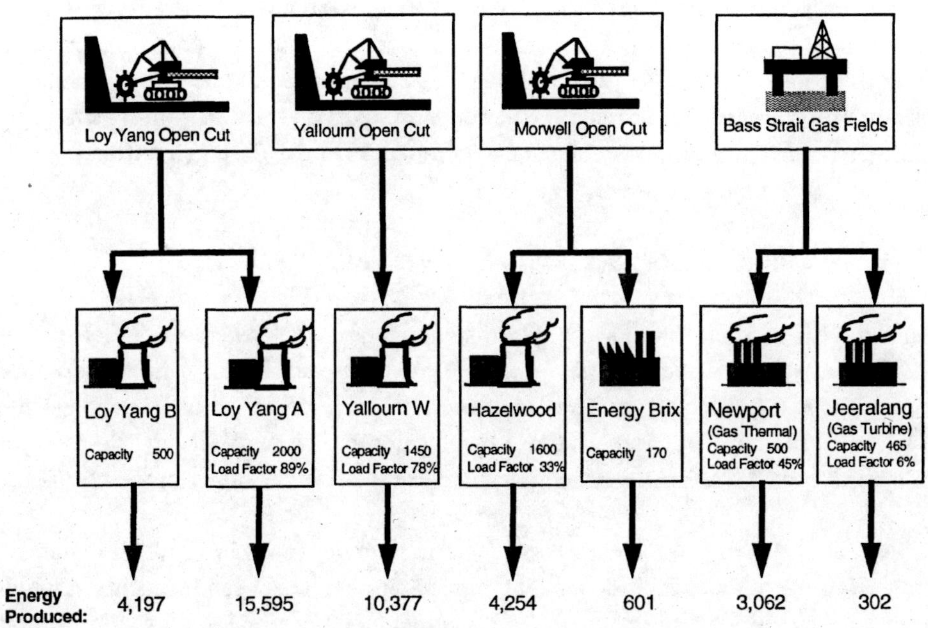


Source: Department of Treasury and Finance

Figure 3: Major brown coal and gas-fired generation.

## THERMAL POWER STATIONS

Most electrical energy in Victoria is generated in brown coal power stations located in LaTrobe Valley approximately 150 km from Melbourne. Victoria has brown coal reserves amounting to more than 42 giga tonnes, or in excess of 1,300 years supply at current production levels. Three open cut brown coal mines operating in the Latrobe Valley - Yallourn, Loy Yang and Morwell - together produce approximately 50 million tonnes of brown coal annually. The mines are operated by the three largest brown coal-fired power stations - Yallourn W, Hazelwood and Loy Yang A [5].

**GAS POWER STATIONS**

Two gas power stations are owned and operated by *Generation Victoria*: (trading as Ecogen Energy) owns and operates two gas-fired stations:

*Newport D* (500 MW) in Williamstown, near the Melbourne central business district (Thermal power unit)
*Jeeralang* (466 MW) in the Latrobe Valley (Seven gas turbine units)

GASCOR currently supplies gas to the two plants, but Generation Victoria is free to purchase gas on commercial terms from any supplier. Historically, these power stations have operated in intermediate and peaking roles, but this role has come under great pressure in the competitive market due to improved performance of base load generating plant. Future utilisation of the gas plants will depend in part on the terms on which gas is bought, as gas is currently relatively expensive compared to coal-fired generation [5].

## 5.3 LOCAL HYDRO POWER STATIONS

*Southern Hydro* was formed on the disaggregation of Generation Victoria. It comprises three hydro-electric schemes: Kiewa, Dartmouth and Eildon, and some minor generating plant at other locations. The *Dartmouth* scheme, in the Mitta Valley, and the *Eildon* scheme, in the Goulburn Valley, are located at large reservoir storages in Victoria's north-east. The *Kiewa* scheme is situated in the Kiewa Valley in the north-east of Victoria.

The primary purpose of each storage is to provide irrigation. Consequently, generation from each scheme is dependent on irrigation release requirements, though negotiations are underway to make the arrangements more flexible for commercial generation operations.

Unlike Dartmouth and Eildon, the Kiewa scheme has minimal requirements to release water imposed on it by an external authority, and operates primarily for electricity generation. As a 'run of river' scheme, its only water release constraint is the size of its storage capacities. This constraint, however, means that the scheme must release water more or less within the same season it is received.

Southern Hydro was sold by the State Government in November 1997.

### SNOWY MOUNTAIN HYDRO ELECTRIC SCHEME

The Snowy Mountains Hydro-Electric Authority (SMHEA) was constructed after World War II. The primary purposes of the scheme are to divert water to the Murray and Murrumbidgee river systems for irrigation and river management, and to produce renewable electricity for South-Eastern Australia through an integrated network of dams, tunnels, pump storage and power stations.

The Scheme's operational area covers some 3,200 square kilometres, wholly within the Kosciusko National Park in Southern NSW. Currently, the Scheme has a total generating capacity of 3,746 MW, and produces an average of 5,000 GWh per year.

From October 1994 until May 1997, the 29% Victorian entitlement was bid into the competitive market by the SECV, trading as the Snowy Trader. The Trader actively bids the output into the Victorian wholesale market and contracts with other retailers and generators on a commercial basis to provide up to 1,500 MW of peaking generation into Victoria.

Since the commencement of the first stage of the NEM in May 1997, the three Governments have merged their electricity entitlements into Snowy Hydro Trading Limited. This is a prelude to full corporatisation in 1998 and ensures that Snowy is not disadvantaged in the increasingly competitive national market [5].

### INTERCONNECTION TRADER

NSW and Victoria had surplus generation capacity while South Australia was forecasting supply shortages, so it appeared that a trading agreement would be mutually beneficial. The Interconnection Operating Agreement (IOA) was signed by the three State-owned electricity utilities to provide for interstate transfers of electricity from 1990 to 2010. Since October 1994, the IOA Trader has acted effectively as both a customer and a generator into the Victorian wholesale market at various times, but with the advent of NEM1, the States agreed to terminate the IOA and the SECV was able to terminate the IOA Trader [5].

### CAPACITY OF POWER STATIONS

Table 1 summarises the generation capacities of Victoria's major power stations, plus other Victorian entitlements.

Table 1. Victorian generator capacity [5].

| *Power Station* | *Nameplate Rating (MW)* | *Installed Plant (MW)* | *Energy Generated(1) (GWh)* | *Capacity Factor(2) (%)* |
|---|---|---|---|---|
| ***Brown Coal-fired:*** | | | | |
| *Loy Yang A* | *2,000* | *4 x 500* | *15,595* | *89* |
| *Hazelwood* | *1,600* | *8 x 200* | *4,254* | *30* |
| *Yallourn W* | *1,450* | *2 x 350*<br>*2 x 375* | *10,377* | *82* |
| *Loy Yang B (3)* | *1,000* | *2 x 500* | *4,197* | *96* |
| *Energy Brix* | *170* | *1 x 60*<br>*3 x 30*<br>*1 x 20* | *601* | *40* |
| *Anglesea* | *150* | *1 x 150* | *1,030* | *78* |
| ***Natural Gas-Fired:*** | | | | |
| *Newport D* | *500* | *1 x 500* | *3,062* | *70* |
| *Jeeralang* | *466* | *3 x 80*<br>*4 x 56.5* | *302* | *7* |
| ***Hydro-Electric:*** | | | | |
| *Kiewa Group* | *184* | *2 x 12*<br>*4 x 15.4*<br>*6 x 18* | *307* | *19* |
| *Dartmouth* | *150* | *1 x 150* | *300* | *46* |
| *Eildon/Rubicon/Cairn Curran* | *135* | *2 x 60*<br>*6 x 4.6 or less* | *77* | *12* |
| ***Other Entitlements:*** | | | | |
| *Snowy* | *1,084* | *Various* | *1,319* | |
| *Hume* | *25* | | *77* | |
| *Total* | *8,414* | | *41,860* | |

(1) Energy generated during the year ended 30 June 1996.

(2 ) For the year ended 30 June 1996. Calculated as energy generated (in MWh) divided by (nameplate rating (MW) multiplied by 8,760 hours).

(3) Loy Yang B2 was commissioned during this period and generated for approximately six months.

Source: Victorian Power Exchange.

## 6 DISTRIBUTION AND RETAIL BUSINESSES

There are five large energy retailers covering the area of the State of Victoria.

***Powercor*** is a distribution company that supplies electrical energy to the Western parts of Victoria, counting for 55% of geographical part of the state. The company has over 500,000 customers covering 150,000 sq. km of supply area. Powercor owns and maintains over 100,000 km of distribution lines selling 7350 GWh electrical energy per year.

***Solaris*** supplies Melbourne's North-Western suburbs. It has over 230,000 customers including 4700 industrial, 19,000 commercial and over 210,000 domestic. Solaris owns and maintains 4000 km of supply lines. Selling 3,324 GWh electrical energy per year.

***City Power*** supplies the central area of Melbourne including CBD covering 157 square km. It owns and maintains 4000 km of supplying lines, mostly underground cables.

***United Energy*** supplies the Southern part of Melbourne covering an area of 1,514 sq. km. It serves approximately 525,000 customers. United Energy owns and maintains over 16,000 km of supply lines.

***Eastern Energy*** supplies electrical power to the eastern part of Victoria covering an area over 8,000-sq. km. It serves 474,000 customers. Eastern Energy owns and maintains over 55,000 km of supply lines selling 4904 MWh of electrical energy per year.

The five distribution companies obtained retailer licences on 3 October 1994. Other retailers who do not own distribution networks are listed below with the dates of the licences issue [6].

| | |
|---|---|
| EnergyAustralia | 17 August 1995 |
| Energy Victoria Ltd | September 1995 |
| Boral Energy (VIC) Pty. Ltd. | 22 November 1995 |
| Yallourn Energy Pty Ltd | 13 May 1996 |
| QTSC (trading as ERGON Energy) | 4 June 1996 |
| ETSA Power Corporation (Victoria) Pty Ltd | 30 July 1996 |
| Great Southern Energy Victoria Pty Ltd | 25 February 1997 |
| Advance Energy | September 1997 |
| Kinetik Energy | January 1998 |
| Energy 21 | January 1998 |

## REFERENCES

1. Introducing The Victorian Power Exchange. *Vpx 101.1*, July 1995, Melbourne.
2. Vicpool. Victoria's Competitive Electricity Market. *Vpx 501.2*, May 1995, Melbourne.
3. Vicpool Metering. *Vpx 601.0*, November 1994, Melbourne.
4. The Victorian Power System Control Centre. *Vpx 801.1*, October 1995, Melbourne.
5. *Victoria's Electricity Supply Industry, Towards 2000*, June 1997. Energy Project Division, The State Government of Victoria.
6. *Licences*. Office of the Regulator-General. 1998.

# PART 2 MARKET REGULATION

****** 4 ******

# CODES AND RULES IN COMPETITIVE ELECTRICITY MARKETS

**WLADYSLAW MIELCZARSKI AND GRAZYNA MICHALIK**
Monash University, Department of Electrical and Computer Systems Engineering
Melbourne, Australia, Email: Wlad@basil.eng.monash.edu.au

**Summary:** The paper presents Code and Rules, which regulate the Victorian open electricity market. These regulations have been introduced by the Office of the Regulator-General. The Codes and Rules discussed are: Vicpool Rules, System Code, Wholesale Metering Code, Distribution Code, Supply & Sale Code and Retail Tariff Metering Code. The regulations were introduced after a consultation process involving all interested parties and public. The National Electricity Market introduced in May 1997 is administrated using the National Electricity Code.

## 1 INTRODUCTION

An open electricity market in Victoria is regulated by Rules and Codes introduced by the State Government institution: the Office of the Regulator-General. All pool participants have to obey Codes and Regulations. This is required by a license issued by the Government. The following Codes and Pool Rules have been developed:

- VicPool Rules;
- System Code;
- Wholesale Metering Code;
- Distribution Code;
- Supply & Sale Code; and
- Retail Tariff Metering Code.

The Codes are to be read in conjunction with other instruments, for example:

- the State Electricity Commission Wiring Regulations 1992;
- the State Electricity Commission (Code of Practice for Tree Clearing) Regulations 1991; and
- the Electrical Safety Code of Practice.

## 2 POOL RULES

The behaviour of participants in Victorian Wholesale Electricity Market (VicPool) is regulated by the Pool Rules [1]. The Rules cover areas including the application and admission of participants, prudential requirements, dispute resolution, bidding into the Pool, generator scheduling, determination of payments receivable by Generators, customer charges and the payment of Pool fees. The Pool Rules regulate the operations of the Wholesale Market and are binding on all licensees.

## 3 SYSTEM CODE

The purpose of the System Code is to set out requirements [2]:

- to ensure the safe and secure operation of the Victorian power system;
- relating to the operation of the Victorian electricity Transmission System and equipment connected to the Victorian electricity Transmission System;
- for design and technical matters relating to connection to the Victorian electricity Transmission System; and
- for quality of supply standards which apply to points of connection to the Victorian electricity Transmission System.

The System Code is binding on all licensees. It sets out procedures for scheduling the dispatch of centrally dispatched Units in accordance with the Pool Rules, performance requirements placed on Distributors, Generators and PowerNet Victoria, the company which owns the transmission network to ensure adequate technical performance of the interconnected power system and other matters relating to the operation of the system as a whole.

## 4 WHOLESALE METERING CODE

The purpose of the Wholesale Metering Code is to ensure that the amount of electricity transferred between Pool participants is measured to facilitate trading through the Pool [3]. The Wholesale Metering Code sets out the obligations for the various industry participants in relation to the timely provision, testing, operation and maintenance of appropriate metering equipment. Provision is also made for situations where metering equipment is not operating correctly.

## 5 Distribution Code

The Distribution Code is designed to regulate in a safe, efficient and reliable manner [4]:

- the supply of electricity from a distribution company's Distribution System; and
- the way in which a customer's electrical installations affect the Distribution System from which a supply of electricity is received.

The distribution licenses require each Distribution Company to comply with the Distribution Code. As a result of requirements imposed on,

- each Distribution Company under its distribution and retail licenses; and
- independent Retailers, under their retail licenses,

Each supply and sales contract involving either a franchise or non-franchise Customer requires the relevant Customer to comply with those provisions of the Distribution Code which are expressed to impose obligations on Customers. This approach has been taken so that it is possible to regulate the way in which a customer's electrical installations affect Distribution Systems.

## 6 Supply and Sale Code

The purpose of the Supply and Sale Code is to set the conditions under which a Distribution/Retailer Company sells electricity to franchise Customers [5]. It is through their retail licenses that a Distribution/Retail company is compelled to comply with the Supply and Sale Code.

Under the Code, the Distribution Companies are obliged to provide and maintain equipment for the supply of electricity up to the point of supply, to provide and maintain standard metering and necessary ancillary equipment, to make all reasonable efforts to provide supply at the standard voltage and frequency, not to disconnect supply to franchise customer's premises except as detailed in the Code, and to be sensitive to the environment and the amenity of the area (but reserving the right to erect overhead electric lines, underground electric lines or substations).

The obligations of franchise Customers are also prescribed. The Code deals with billing, payment and disconnection, tariff rates, confidentiality of franchise customer information, supply quality standards and other matters relating to the services to franchise Customers. The Supply and Sale Code is being reviewed by the Office's Customer Consultative Committee in consultation with the Distribution Companies.

## 7 Retail Tariff Metering Code

The Retail Tariff Metering Code regulates the basis for installation of new metering equipment designed to measure and record the amount of electricity supplied from a Distribution System to a customer's electrical installation and the operation and maintenance of existing and new metering equipment [6]. This Code establishes rights and obligations in respect of metered data.

It is a term of both distribution and retail licenses that the license holder complies with the Retail Tariff Metering Code. Compliance with the Retail Tariff Metering Code by franchise Customers is a term of the Supply and Sale Code. In so far as contract customers are concerned, a supply and sales contract between,

- a Distribution Company and a franchise or contestable contract Customer, or
- an independent retailer and contestable Customer,

## 8 Implementation

Each of the licenses issued by the Office requires the licenses to observe certain Codes and Pool Rules. All licensees listed must comply with [7],

- the Pool Rules;
- the System Code; and
- the Wholesale Metering Code.

The System Code also applies to Extra High Voltage (EHV) Customers, if the Office of the Regulator General so directs.

Each distribution and retail licensee must also comply with the following Codes:

- the Distribution Code;
- the Supply and Sale Code; and
- the Retail Tariff Metering Code.

The Distribution Code also applies to embedded Generators and Customers. The Supply and Sale Code also applies to Customers; the Retail Tariff Metering Code also applies to each franchise Customer and, if required by a contract, to a non-franchise Customer.

## 9 HIERARCHY

To the extent that the Pool Rules are inconsistent with the provision of any Code, the Pool Rules prevail. The hierarchy of the Codes is as follows [7]:

- the System Code;
- the Wholesale Metering Code;
- the Distribution Code;
- the Supply and Sale Code; and
- the Retail Tariff Metering Code.

That is, the extent that a Code is inconsistent with the provision of another Code below it in the above list, the first mentioned Code in the above hierarchy generally prevails.

## 10 CUSTOMER CONSULTATIVE COMMITTEE

Because Codes and Rules are public documents affecting all pool Participants and Customers, their introduction is preceded by public discussion and consultation involving many organisations. From the legal point of view, Codes and Rules are introduced by the State Government which branch, The Office of the Regulator General, prepares a draft written with the assistant of independent experts, like the authors of this report. The Office of the Regulator-General is assisted in its consideration of Customer issues by a Customer Consultative Committee which it has established [7].

*Membership* - Committee's membership is comprised of representatives of each of the following organisations:

- Australian Chamber of Manufactures;
- Australian Federation of Consumer Organisations;
- Building Owners and Managers Association;
- Consumer Advocacy and Financial Counselling Association;
- Environment Victoria;
- Victorian Employer's Chamber of Commerce and Industry;
- Victorian Council of Social Service; and
- Victorian Farmers Federation.

*Observers* - Observers on the Committee include a representative from the following departments and agencies:

- the Department of Agriculture, Energy & Minerals;
- the Department of Conservation & Natural Resources;
- the Department of Health & Community Services;
- the Environment Protection Authority; and
- the Office of Fair Trading & Business Affairs.

The Committee's role is,

- to provide a forum in which information relevant to the achievement of the Committees objectives may be disseminated; and
- customers' representatives may exchange views on customers' entitlements, issues and needs.

The Committee's role is also to advice the Office of the Regulator-General of customer issues within the objectives of the Office of the Regulator-General and, in particular, to provide,

- advice on the needs of industrial, commercial and residential customers generally; and
- customers' views on particular issues - referred to the Committee by the Office of the Regulator-General or agreed by the Committee.

The Committee has established Working Groups to, amongst other things:

- review the electricity industry's Supply & Sale Code and the water industry's customer contract which forms part of the water industry licensing regime;
- examine issues relating to access to essential services, security deposits and disconnections; and
- focus on the development of appropriate social indicators and customer service standards reporting procedure.

Although the process of consultation seems to be quite complicated, it is necessary to allow all interested parties to express their opinions and concerns prior to introduction of any rules.

## 11 THE NATIONAL ELECTRICITY CODE

The National Electricity Market, which was introduced in May 1997, is regulated by the National Electricity Code. The code has its genesis in the Special State Premiers' Conference in October 1990 and July 1991, which led to the formation of the National Grid Management Council. The council published in October 1993 the discussion paper in which it recommended a range of regulatory arrangements for the national electricity grid. It is indicated that the operation of the national grid will be regulated by way of Code of Conduct [8].

The development of the national grid and the Code is overseen by a national regulator such as the Australian Competition and Consumers Commission and the states regulatory bodies like the Office of the Regulator-General in Victoria to be consistent with the state regulations. The States of New South Wales (NSW), Victoria, Queensland, South Australia and the Australian Capital Territory have agreed to the enhancement of cooperative legislation for the implementation of the regulatory arrangement. This legislation is called the National Electricity Law. Two institutions have been established: the National Electricity Code Administrator (NECA) and the National Electricity Market Management Company to develop and introduce the NEC to operation of the National Electricity Market. NECA is the corporate body responsible for administrating the Code. NEMMCO is responsible for operating and administering the market in accordance with the Code.

The objective of the National Electricity Code is to provide:

- regulation to achieve the market objectives,
- a set of rules governing market operation, system security, network connections and network pricing,
- framework to dispute resolutions,
- sanctions in cases of breaches of the Code,
- and processes for changing the Code.

The National Electricity Code is organized in ten chapters entitled as follows [8]:

- Introduction
- Code participants and registration
- Market rules
- Power system security
- Network connection
- Network pricing
- Metering
- Administrative functions
- Jurisdictional derogation and transitional arrangements
- Glossary

Although the Code has been introduced after a long discussion involving all interested parties, regulators, associations and public, it has to be a live document which be able easily incorporate changes resulting from the experience of National Market operation.

## REFERENCES


1 *VicPool Rules*, Office of the Regulator-General, Melbourne, Victoria
2 *System Code*, Office of the Regulator-General, Melbourne, Victoria
3 *Wholesale Metering Code*, Office of the Regulator-General, Melbourne, Victoria
4 *Distribution Code*, Office of the Regulator-General, Melbourne, Victoria
5 *Supply & Sale Code*, Office of the Regulator-General, Melbourne, Victoria
6 *Retail Tariff Metering Code*, Office of the Regulator-General, Melbourne, Victoria
7 *Electricity Industry Regulatory Statement*, Office of the Regulator-General, Melbourne, Victoria
8 *National Electricity Code*, National Electricity Code Administrator, Melbourne/ Sydney

****** 5 ******

# Office of the Regulator-General and its Role in Competitive Electricity Markets

**John Tamblyn**
The Regulator-General, The Office of the Regulator-General, Melbourne, Australia

**Summary:** The Office of the Regulator-General, is the independent economic regulator set up by the State Government of Victoria, Australia to regulate designated industries including electricity, water, grain, ports, gas and railways. The paper presents the Office's structure, its functions and main objectives. The presentation is focused on the Office's activities in regulation of the competitive electricity market in Victoria.

## 1 Introduction

Australian and overseas experience suggests that restructured government business enterprises may abuse their market power at the expense of customers and competition. To avoid such negative impacts and to maximise benefits to the community, some economic regulation is necessary. The Office of the Regulator-General, Victoria has been established to oversee that economic regulatory framework.

The Office of the Regulator-General (the Office) was established on 1 July 1994 under the *Office of the Regulator-General Act 1994* (the Act). The Act established the Office as an economic regulator and lays a foundation for the Office to perform its functions and exercise its powers in respect of regulated industries operating under relevant legislation. The Office's corporate objectives are to give effect to the Government's microeconomic reform agenda by regulating the electricity water, grain handling, ports and other industries which the Government may include within the Office's mandate.

The following industries were regulated within the ambit of the Office by virtue of the relevant legislation specified as follows:

- the electricity industry (from 3 October 1994) - the *Electricity Industry Act 1993;*
- the water industry managed by three Melbourne metropolitan licensees, Melbourne Water Corporation and Melbourne Parks & Waterways (from 1 January 1995) - the *Water Industry Act 1994;*
- the industry of facilitating the export shipping of grain (from 1 July 1995)
- the *Grain Handling and Storage Act 1995;*
- the port industry in the ports of Melbourne, Geelong, Portland and Hastings (from 1 January 1996) - the *Port Services Act 1995;*
- certain rail services (from 1 April 1997) - the *Rail Corporations Act 1996;* and
- the Victorian gas industry (from December 1997) - the *Gas Industry Act 1994*

It is anticipated that the non metropolitan urban water industry will also come within the ambit of the Office.

The relationship between the legislation, the regulations and Orders may be summarised as shown in Figure 1.

As far as reasonably practicable, the regulatory framework is light handed. That is, it lets the participants in the regulated industries get on with their business and provides for intervention only where it is necessary to correct misuse or abuse of market power in the interests of competition or customers.

## 2 The Office's Objectives

In performing its functions and exercising its powers the Office has the following general objectives under the Act:

- to promote competitive market conduct;
- to prevent misuse of monopoly or market powers;
- to facilitate entry into the relevant markets;
- to facilitate efficiency in regulated industries; and
- to ensure that users and consumers benefit from competition and efficiency.

The relevant legislation referred to above sets industry specific objectives for the Office that reflect those in the Act. If there is a conflict between the objectives under the relevant legislation and those in the Act, those in the relevant legislation prevail. That is, the Office must perform its functions and exercise its powers in a manner it considers best achieves the objectives in the relevant legislation.

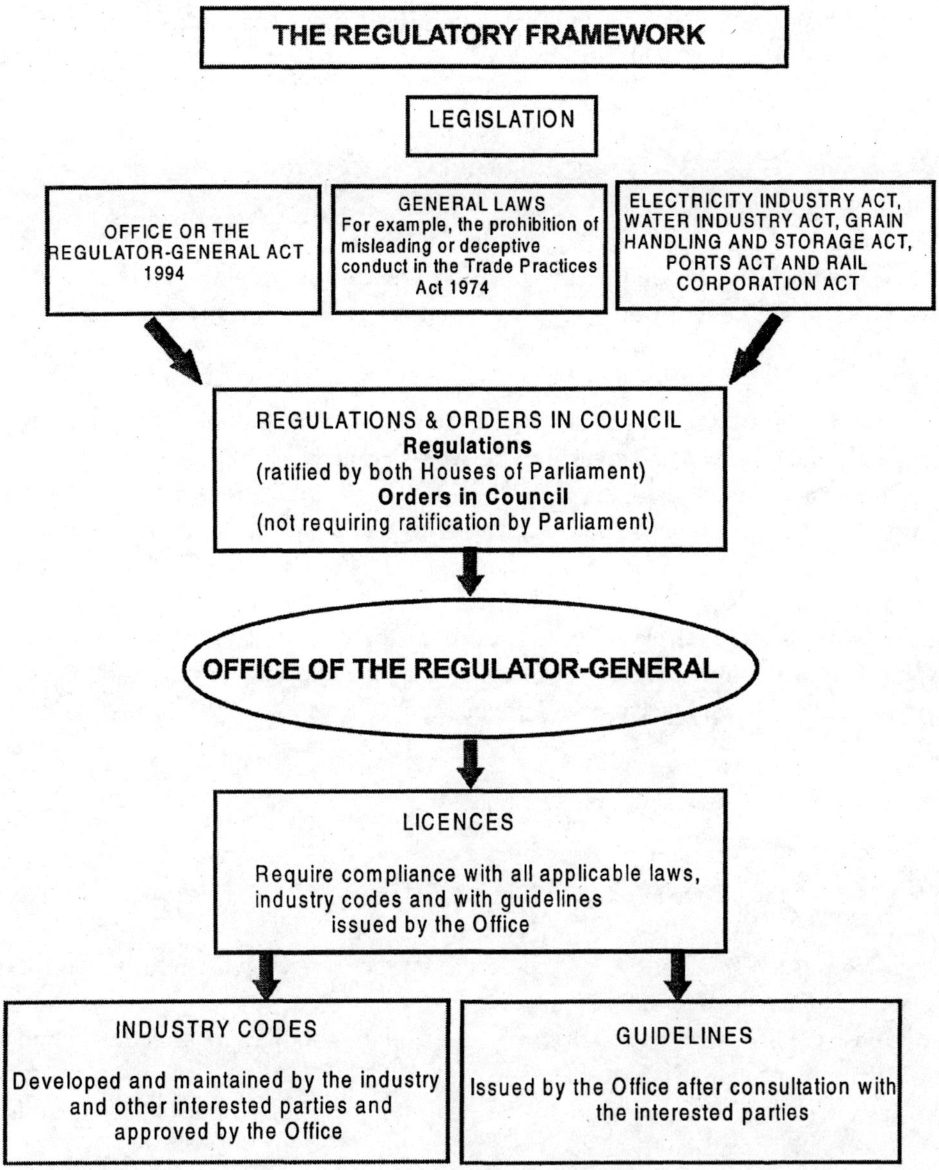


Figure 1 The regulatory framework

Consistent with the Government's light handed approach to regulation, the Office does not have carte blanche to achieve its objectives in any way it sees fit. It is not open to the Office to rely solely on its objectives to promote competition and protect consumers and ignore:

- the carefully delineated powers that enable it to achieve those objectives; and

- the quite detailed procedures that are set out in the Act and the relevant legislation on how the Office is to achieve its objectives.

## 3 Independence of the Office

The thrust of the regulatory framework is that the Office is independent of the government of the day. The independence of the Office was strengthened by the removal of the power in the government to declare a statement of policy and other amendments made to the Act which remove:

- a requirement that the Office perform its functions of administering the Act subject to the general direction and control of the Minister; and
- requirements in section 25 and 26 of the Act that determinations made by the Office under those sections must be consistent with any statement of government policy.

A further safeguard for the independence of the Office is to be found in the Act which provides that the Office is not subject to the direction or control of the Minister in respect of any determination, report or inquiry except as provided in that Act or another Act, such as the relevant legislation referred to above.

## 4 The Office's Functions

The functions of the Office as expressed in the Act are:

- to perform such functions as are conferred on the Office by the relevant legislation under which a regulated industry operates;
- to advise the Minister for Finance on matters relating to the economic regulation and restructuring of regulated industries;
- to conduct inquiries and report to the Minister for Finance on matters relating to regulated industries;
- to advise the Minister for Finance in relation to any matter referred to the Office by the Minister;
- to perform such other functions as are conferred or imposed on the Office by the Act; and
- to administer the Act.

### 4.1 Powers

Both the Act and the relevant legislation referred to above confer powers on the Office relevant to it achieving its objectives. The Office's specific powers under the Act include the power to regulate prescribed prices in respect of prescribed goods or services supplied by or within a regulated industry.

The Act provides that the Office may exercise such powers (including the power to make determinations) for or with respect to:

- standards and conditions of service and supply;
- licensing;
- market conduct; and
- other economic regulatory matters, as may be conferred on the Office by relevant legislation or by an Order-in-Council made under the Act.

To ensure that the Office can effectively carry out its responsibilities, it has significant powers to obtain information from the regulated industries.

## 4.2 Inquiries and Reports

The Office may, after consultation with the Minister for Finance, conduct an inquiry if it considers an inquiry is necessary or desirable for the purposes of carrying out its functions. The Office must also conduct an inquiry into any matter which the Minister may refer to the Office. Further particulars of the procedures that the Office must observe in conducting an inquiry and its general approach to the conduct of an inquiry are contained in its electricity regulatory statements.

## 4.3 Enforcement Powers

The Office has the significant powers to reinforce initiatives it may take to achieve its objectives. Firstly, the Office has the power to include relevant conditions in the licences it issues or administers. Breach of a licence condition exposes the licensee to risk of a significant penalty (an initial $100,000 flagfall and $10,000 per day for a continuing breach). Secondly, a breach of a determination made by the Office may also attract a significant penalty. Thirdly, the Office has significant powers to obtain information, not only from licensees but also from other parties. A penalty attaches to a failure to provide the Office with information or documents ($10,000 or two years imprisonment).

## 4.4 Appeals

A person who is aggrieved by a determination of the Office may appeal against the determination on the grounds that:

- there has been bias; or
- the facts on which the determination is based have been misrepresented.

An appeal must be lodged with the Office within seven working days after the determination has been published and must be heard by an appeal panel, consisting of three persons appointed by the Minister for Finance from a pool of persons appointed by the Governor in Council because of their knowledge of, or experience in, industry, commerce, economics, law or public administration.

An appeal must be heard and decided by the appeal panel within two weeks of the appeal notice being lodged with the Office.

Appeals to the courts challenging a determination or order by the Office are limited to grounds that:

- there was no power to make the determination or order; or
- procedural requirements in relation to the making of the determination or order were not complied with.

## 4.5 Customer Consultative Committee

The Office is assisted in its consideration of customer issues by a Customer Consultative Committee. The Committee's membership is drawn from representatives of the following organizations:

- the Australian Chamber of Manufactures;
- the Building Owners and Managers Association;
- the Financial Consumer Rights Council (Inc);
- the Consumers' Federation of Australia;
- Environment Victoria;
- the Victorian Council of Social Service;
- the Victorian Employers' Chamber of Commerce and Industry; and
- the Victorian Farmers Federation.

The Office is committed to consultation. Since its establishment, it has developed the following customer and industry consultation framework. The Office consultation framework is shown in Figure 2.

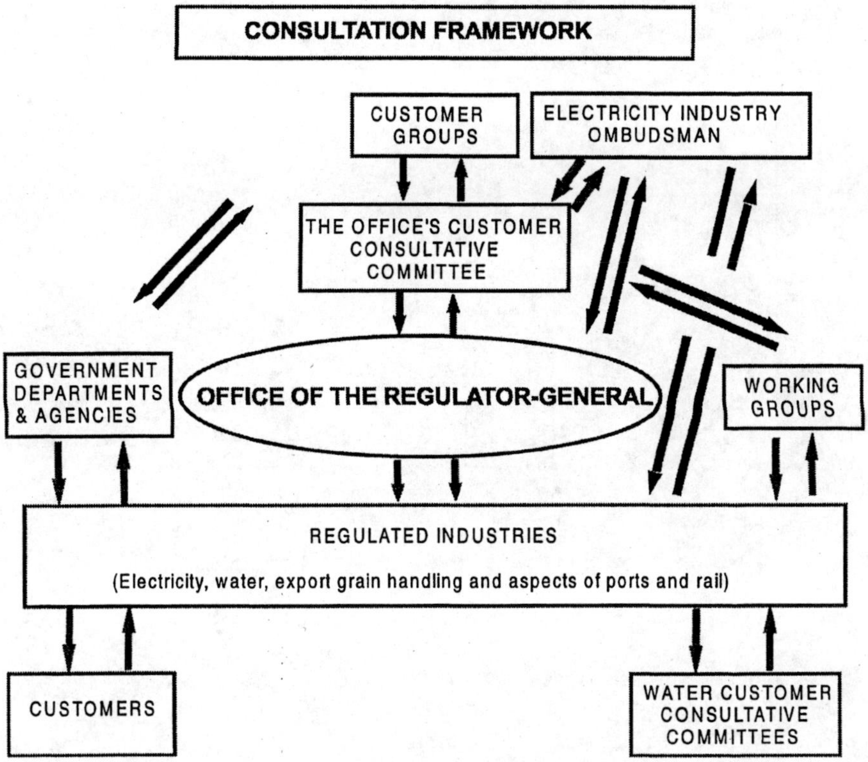


Figure 2 Consultation framework

## 5 THE OFFICE STRUCTURE

The Office consists of one member appointed as the Regulator-General. The Regulator-General is appointed for a five year term by the Governor in Council on the basis of his or her knowledge of, or experience in, one or more of the fields of industry, commerce, economics, law or public administration.

The new Regulator-General is Dr John Tamblyn, a former senior adviser to the Australian Competition and Consumer Commission.

The Minister for Finance may appoint persons as Associate Regulators-General who are qualified for appointment because of their knowledge of, or experience in, industry, commerce, economics, law or public administration. It is envisaged that the Office may recommend to the Minister the appointment of Associate Regulators-General to undertake specific inquiries.

Staff of the Office are appointed or employed under the *Public Sector Management Act 1992*. An outline of the structure of the Office is shown in Figure 3.

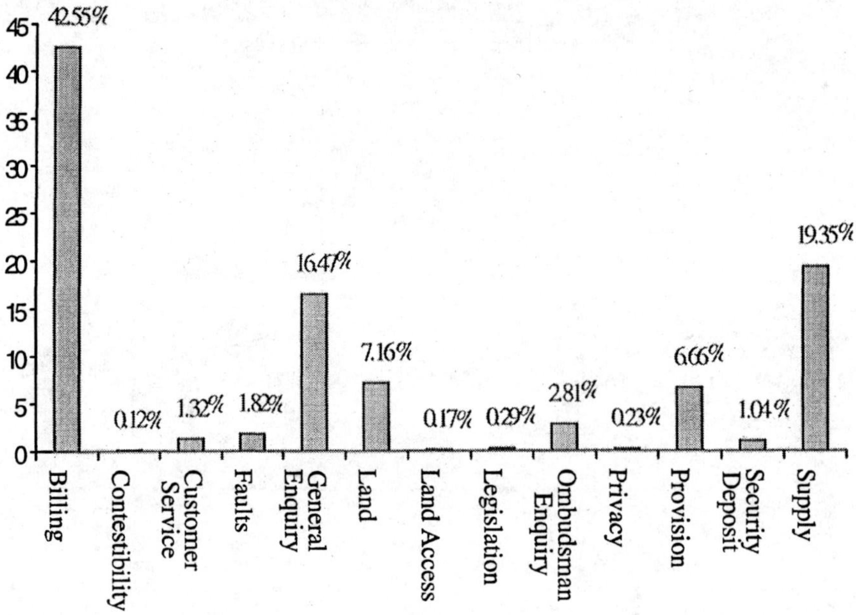


Figure 3 The structure of the Office.

## 6 Major Challenges

Major challenges faced by the Office in the electricity industry over its next three years include:

- the determination in the year 2000 of distribution network charges to apply post-2000 - as outlined below, preparation for this task has commenced with the publication of *Electricity Industry Guideline No. 3: Regulatory Information Requirements;*
- ensuring a smooth and orderly transition to open competition involving, amongst other things, major education programs to enable customers to make an informed choice of supplier and resolving metering versus deemed load profile issues;
- developing and implementing an operational audit regime similar to that developed in relation to the water industry;
- the review of relevant standards and, where appropriate, clarification and restatement of quality and reliability standards; and
- expanding and refining the Office's performance reporting regime.

## 7 Divisional Activities

While the Office's structure over the past three years was appropriate to meet the initial challenges it faced, experience in meeting those challenges suggested the need for realignment. This has been carried out to provide a structure suited to meet the major challenges outlined above as well as in managing the Office's routine continuing tasks.

There are four operational Divisions in the Office - Industry Performance, Pricing, Competition and Market Conduct, and Customer Relations. Each Division works on projects relating to its specialist skills. The expertise of two or more Divisions combines on cross-Divisional projects when required.

Examples of the activities of the Divisions include:

### 1 Revision of Electricity Industry Guideline No 3: Regulatory Information Required

Due to clauses in the electricity distribution /retailer licences the licensee must ensure the separate accounts for each electricity business and provide to the Office the information required. The Division has developed a guideline providing for the collection, allocation and recording of business data.

### 2 Revision of Electricity Industry Guideline No 5: Connection & Use of Systems Agreements

Connection and use of system agreements are an essential ingredient of a successful competitive electricity market. The Division has developed a guideline outlining principles, consistent with the regulatory regime and good electricity practice, against which the Office may decide whether connection or use of system agreements are fair and reasonable.

### 3 Setting Procedures for Distribution Loss Factors

Victorian Power Exchange must allocate distribution losses to second tier customers ie. non-franchisee customers which buys electricity from a retailer other than the host retailer, by applying relevant distribution loss factors listed in a schedule to the Rules. On the other hand, distributors must perform an annual review of the distribution loss factors for their business and submit them to the Pool Consultative Committee by 31 March in each year for approval to take effect from 1 July each year.

Prior to 1 July 1996, the Pool Rules required VPX to calculate loss factors on an average basis at each of four supply/voltage levels. There was no differentiation to take account of variations in losses attributable to the length of a customer's supply line.

In October 1995 VPX convened a Distribution Losses Working Group, consisting of representatives of each distribution business for the purposes of reviewing the distribution loss factors to be applied from 1 July 1996.

The Working Group's output consisted of a paper containing two tables of distribution loss factors - one showing distribution loss factors averaged across all customers at each connection voltage level and the other showing loss factors for short and long sub-transmission lines and high voltage distribution lines.

The Working Group's paper was submitted to the Pool Consultative Committee which voted 8 to 1 (with 1 abstention) in support of the de-averaged loss factor alternative. The necessity to consult with interested parties delayed the Office's decision whether to approve the amendments until October 1996 when it amended the Pool Rules to allow for the introduction of de-averaged distribution loss factors with effect from 1 July1996.

In January 1997, one of Distribution Businesses, Powercor, issued a Supreme Court challenge to the Office's amendment of the Pool Rules. As the Office understands it, Powercor does not challenge the merit of the Office's decision but focuses on the Office's procedures and, in particular, the Office's power to make the amendment effective from 1 July 1996. The Office is defending the challenge.

## 4 VICPOOL MONITORING

There is a risk that unless the Office is ever vigilant, pool participants may seek to *game* the pool (ie - take advantage of unintended consequences of Pool Rules) or, even worse, act in concert to fix pool prices.

To minimise such risk, the Office has established a protocol with Victorian Power Exchange (VPX) which provides for VPX to monitor, and report to the Office on, the behaviour of pool participants. The protocol describes:

- matters to be monitored by VPX relating to the competitive efficiency of the wholesale market for electricity including, but not limited to, four areas of concern, namely, participant behaviour, market conduct, market design adequacy and wholesale market outcome;
- ranges and levels of activities which, if exceeded, will require VPX to notify the Office accordingly.

The monitoring is managed by a *Pool Monitoring Protocol Working Group* with membership consisting of representatives from the Office and VPX. It is proposed that responsibility for the *Pool Monitoring Protocol Working Group* will transfer to the National Electricity Code Administrator on the commencement of the national electricity market.

## 5 INVESTIGATION OF INCIDENTS

The office is also involved in investigation on causes of some incidents. One incident, known as the "AFL Park blackout" involved the halting of a night football match when the electricity supply was interrupted. The second incident resulted in a power surge damaging several properties when a 600V, dc tram line cable came into contact with a 240V power line.

A number of issues emerged from the Office's investigation of this and other incidents which will be addressed in the Office's current review of the Distribution Code. The Distribution Code contains, amongst other things, standards which distribution businesses must observe in supplying electricity to its customers.

## 6 HALF HOURLY METERING CHARGES

The Office published its decision on half hourly metering charges for 1996-97 in October 1996. The decision was based on advice from an independent consultant and from extensive consultation with the distribution businesses and the Victorian Power Exchange.

The report allowed the Office to identify variables within the major drivers which have significant impact on the costs:

- the standard of the meters for 1 MW+ and 750MWh/year customers (a meter built to the standard applicable to a 1 MW+ customer's needs costs approximately $600 more than a meter built to the standard applicable to a 750MWh/year customer);
- labour rates (the labour rates submitted by the businesses ranged from $50 - $79 per hour);
- the number of personnel and time taken to install the meters (figures submitted by the businesses ranged from 2 to 4 people and from 1 to 6 hours per person);
- telecommunications equipment costs (which vary according to whether a fixed or mobile link is used and whether a high voltage line isolation unit is required).

The decision on 1996-97 half hourly metering charges represents a reduction on 1995-96 charges and suggests a downward trend, as at 30 June 1997. The Office will not determine half hourly metering charges for 1997-98 due to the introduction of contestablility in metering services..

## 7 METERING SOLUTIONS FOR 160 MWH/YEAR CUSTOMERS

The next tranche of customers to choose from whom they acquire their electricity, the 160MWh/year customer tranche, become eligible to exercise that choice on 1 July 1998. Where such a customer chooses to buy electricity from a retailer other than the retail arm of its host distribution business, an issue arises as to how best to measure that customer's load so that it may be netted off payments made by the host distribution business to the market operator and charged to the appropriate retailer.

The Office has contributed actively to the development of settlement and metering requirements for the 160-750MWhr/Yr tranche, which is scheduled to open after the National Electricity market commences.

## 8 ELECTRICITY PERFORMANCE REPORTS

In June 1997, the Office released the fourth in a series of reports aimed at encouraging *competition by comparison* amongst the five electricity distribution businesses (CitiPower, Eastern Energy, Powercor, Solaris and United Energy). The report provides a business by business comparison against key customer service indicators, assesses their performance over the period July to December 1996. The next report is due in July, 1998

## 9 RETAILER OF LAST RESORT

During 1997, the *Electricity Industry Act 1993* was amended to empower the Office to vary the licences issued to the five distribution businesses to require them, on terms and conditions approved by the Office, to supply and sell electricity to emergency customers. Emergency customer means a non-franchise customer whose retailer of choice has been suspended from trading in the wholesale electricity market.

The amendment provides that in determining the terms and conditions, the Office must have regard to the risks and costs associated with the requirement. The obligation of the licensee to supply and sell electricity to an emergency customer ends after three months or when the emergency customer advises the licensee in writing that the supply and sale is no longer required.

The objective of the retailer of last resort scheme is to ensure to the maximum extent practicable that any disruption to electricity customers is minimised and that where appropriate the retailer of last resort will endeavour to settle any financial consequences of the disruption directly with the suspended retailer.

## 10 SUPPLY AND SALE CODE

A major review and revision of the electricity industry Supply and Sale Code has been completed. The Supply and Sale Code is the customer contract which *franchise customers* (ie - those customers not yet eligible to choose their electricity supplier) have with their distribution business.

The review was undertaken in conjunction with the Office's Customer Consultative Committee, the distribution businesses that supply franchise customers and other interested parties. The revised Supply and Sale Code became effective from 1 February 1997.

The revised Code is a significant improvement on its predecessor, setting out clearly the rights and obligations of both the distribution businesses and franchise customers.

## 11 Electricity Industry Customer Charter

The Supply and Sale Code provides that a distribution business must send to each of its customers a *customer charter* in a form approved by the Office. The charter is a summary of the rights and obligations under the Supply and Sale Code and other key codes and rules. It is a "first" for Victoria, and indeed, Australia.

The Office developed the concept of the customer charter during the Code review process, to meet the information needs of customers. To ensure that the charters sent to customers by the distribution businesses contained a minimum common level of key information, the Office developed a *Benchmark Customer Charter.*

Each distribution business developed its own charter based on the Office's benchmark and submitted it to the Office for approval before it was mailed to all residential and eligible business customers from April to June, 1997.

## 12 Trade Practices Act - Implied Conditions and Warranties

The Office has been working closely with the Australian Competition and Consumer Commission (the ACCC) on the issue whether the *Trade Practices Act's* implied conditions and warranties apply to the contract between a *consumer* (as defined in that Act) and an electricity supplier. The Regulator-General and Office staff have assisted Commission staff understand the issues.

On the basis of legal advice on the application of the *Trade Practices Act's* implied conditions and warranties, the Office and the ACCC take the view that an electricity supplier is liable to compensate *consumers* for damage caused by power surges irrespective of the cause of the surge.

This is a significant breakthrough - many power surges are caused by events beyond a supplier's control, such as lightning strikes or cars hitting power poles. While in the past the suppliers have paid for damage caused by their negligence, they have resisted paying for damage due to causes outside their control.

## 13 Customer Information

Cognisant of the importance of customer information as a market force, the Office has expanded its customer information activities. Its activities in this regard include:

- the development of an Internet site;
- the production of a range of attractive and easy to read customer information brochures; and

- the wide distribution of the brochures through community groups and Members of Parliament.

The Office has also arranged, through the Braille & Talking Book Library, to have its fact sheets, the electricity industry Supply & Sale Code and the Office's Benchmark Electricity Customer Charter recorded on audio tape for customers who are sight impaired.

## 14 Electricity Access to Distribution Networks and Customer Initiated Augmentation

To ensure compliance with the Office's Electricity Industry Guidelines No.1 and 2, (*Access to and Use of Distribution Systems, Statements of Approved Charges and Charges for other Services* and *Distribution Systems Augmentation, respectively*) each distribution business is required to:

- publish a statement of the policies, practices and procedures the business applies in relation to persons seeking access to and use of its distribution network, and
- provide the office with particulars of its augmentation work in order that the Office may monitor the extent to which work is carried out by the business or by third party contractors on behalf of customers .

The Office is responsible for monitoring each distribution business' compliance with the guidelines.

## 15 Amendment of Electricity Industry Guideline No. 2: Distribution Systems Augmentation

The licences issued to the five distribution businesses provide safeguards against a business abusing its de facto monopoly over its distribution system by requiring, amongst other things, that as a general rule the business should:

- call tenders for augmentation work; and
- in calling such tenders, comply with guidelines published by the Office.

The relevant guideline is *Electricity Industry Guideline No. 2: Distribution Systems Augmentation.* As a result of the consultation a number of changes were effected to the guideline in April 1997, including:

- the definition of augmentation;

- stipulating that in determining whether a business must call a tender in relation to an offer to connect, regard should be had to the amount the customer is required to contribute, not the total cost of the augmentation work; and
- variations to the quarterly reporting requirements.

## 16 Mutual Recognition of Electricity Licences

The move of generation, transmission and the wholesale market responsibilities to the national level with the State jurisdictions retaining responsibility for distribution and retail, involves a risk of regulatory duplication and inconsistency. Steps will therefore need to be taken to harmonise as far as possible the various State regulatory requirements. Industry participants have expressed concern about the number of regulators and their different regulatory requirements.

Anticipating such concerns, the Office has, in consultation with the Victorian Department of Treasury and Finance's Energy Projects Division and the New South Wales Treasury, been addressing the issue of how the two States might best give mutual recognition to each State's retail licences. Among other things, legislative change is likely to be required in order to achieve such mutual recognition.

To this end, the Office is developing an issue paper on mutual recognition of retail licences, which will form the basis for further consultation with interested parties.

It needs to be noted that the Queensland department of Mines and Energy, as that State's regulator, is already recognising retail licences issued in Victoria and New South Wales, however at this stage recognition is not mutual.

## 17 Gas Issues Paper

In December 1997, the Office was given the responsibility of regulating gas distribution in Victoria. The *Office of the Regulator-General Act 1994* (as amended), together with the *Gas Industry Act 1994* (as amended) provide the legal basis for the Office to exercise its powers in regulating the gas distribution industry in Victoria.

One of the primary objectives of the Office in gas distribution is to administer an access Code (*Victorian Third Party Access Code for Natural Gas Pipeline Systems*) providing a right of access to services provided by means of distribution pipelines on fair and reasonable terms and conditions.

In November 1997, three gas distributors, Multinet, Stratus and Westar submitted *Access Arrangements* to the Office to be assessed against the requirements of the Code. The Office advertised for public submissions in December 1997, and also released an Issues Paper to guide these submissions. A second round of public consultation took place in March 1998, after the gas distributors had supplemented the information contained in their initial submissions.

The Office expects to issue a *Draft Decision* in May 1998, when it will advertise for further submissions from the public. The *Final Decision* is expected in July 1998.

### 18 INFORMATION TECHNOLOGY STRATEGIC REVIEW

The Office formed an information technology review team to conduct a strategic review of the its information technology environment, framework, operations and requirements. The key deliverables of the information technology strategic review were the development of an Information Technology Strategic Framework and an Information Technology Operating Plan for 1997-98. The main features of the Information Technology Strategic Framework are:

- the documentation of the Office's information technology general principals, minimum standards, operating environment and strategies;
- a structural framework for the delivery of information technology services to the Office; and
- a framework for the preparation of annual Information Technology Operating Plans for the Office.

### 19 LONG TERM STRATEGIC PLANNING AND REGULATORY PLAN

In drafting the 1997-98 Regulatory Plan, the office reviewed its strategic framework for the structure and delivery of its regulatory responsibilities.

The Office's Regulatory Plan represents its detailed annual work plan, which is consistent with the adoption of output based reporting by the government sector.

## 8 TOWARDS THE NATIONAL ELECTRICITY MARKET

In 1997, the national electricity market advanced to what is known as NEM1 Stage 1. NEM1 Stage 1, which commenced on 4 May 1997, involves the initial harmonisation of Victoria's and New South Wales' respective electricity markets to give pool price outcomes substantially similar to a full national electricity market. South Australia participates in NEM1 Stage 1 via the Victorian wholesale market.

Moves towards the National Electricity Market involves the transfer of regulatory responsibility for generation, transmission and the wholesale market from State based regulators, such as the Office, to the Australian Competition and Consumer Commission. The State based regulators will retain responsibility for distribution and retail aspects of the electricity market.

The Competition Principles Agreement between the Commonwealth, the States and the Territories recognises a continuing role for State economic regulation of restructured government business enterprises. Implementation of this regulation will be a continuing role for the Office.

****** 6 ******

# The Electricity Industry Ombudsman, Victoria (EIOV) a Key Plank in the Restructure of the Victorian Electricity Industry

**Fiona McLeod**
Electricity Industry Ombudsman (Victoria) Limited, Melbourne, Australia

## Introduction

Reform of the Electricity Industry within Victoria has been moving at a rapid pace, and the way in which services are provided and by whom is part of the new thinking about many of the ways things are done, a new thinking that could significantly benefit Victoria's electricity users.

The Electricity Industry Ombudsman, Victoria scheme, or EIOV as I will call it, is one of the key linkages between this new thinking and the achievement of a range of positive outcomes.

As an effective alternative dispute resolution agency the scheme has the potential to use its first hand experience to turn the attention of the power companies to major customer service issues and to monitor their success.

Full deregulation of the Victorian electricity market will occur on 1 January 2001. So far, 3 of the 5 market segments have become contestable (non franchise), the fourth arrives on 1 July 1998, and the sale of the transmission system has recently been completed.

The EIOV provides a good case study in how industry dispute schemes can play a positive role in ensuring that quality customer service is provided, be it by company staff, or their contractors and agents. The scheme is available to both contestable and franchise customers, but will be of more benefit to small business and domestic consumers whose claims are likely to come within the scheme's compensation limits of

$10,000 and $50,000 (only with the consent of all parties). The scheme has received a number of cases from contestable customers and some interesting issues arise when we review the contract between the customer and their retailer.

The EIOV is an alternative to formal legal processes for solving complaints. Alternative dispute resolution has much to commend it:

It involves the parties rather than confusing and distancing them as the legal system can do.
It can assist the parties to understand the issues.
It produces a sense of ownership of the outcome and a commitment to its implementation.
It is free and its processes are comparatively speedy.
In terms of its potential impact on customer service, its impartiality means that the data it collects can be an invaluable source of independent information.

There were several industry schemes in existence that served as models for the EIOV. Within the banking and telecommunications industries, industry ombudsman schemes were meeting the needs of customers; their experience was a useful basis for the preparatory work in the development of the EIOV.

One thing was clear from their experience: support for an Ombudsman scheme ought to be a condition of the licence of the privatised electricity distributors, transmitters and retailers within Victoria. There was a second lesson to be learned from the experience of other Ombudsman schemes. The EIOV scheme had to be independent not only in appearance, but it had to be independent in fact.

The structure of the scheme is firmly based upon the principle of independence.

## 2 The Structure

The EIOV is formed as a private company limited by guarantee. The five privately owned distribution businesses CitiPower, Eastern Energy, Powercor, Solaris Power and United Energy are founding members of the EIOV scheme. The transmission company - GPU PowerNet - is also a member of the scheme. The five (5) distribution businesses also hold retail licences. We are currently negotiating with the new independent retailers trading in Victoria for their entry into the scheme. The scheme is established by a Memorandum and Articles of Association which define the three components of the scheme:

1. An independent Council
2. A Board of Directors
3. The Ombudsman.

The independent **Council** is the intermediary between the Ombudsman and the Board. It is composed of an equal number of electricity company representatives and customer interest representatives. Council is headed by an independent Chairman who was appointed on the recommendation of the Regulator-General and after substantial community consultation. The Chairman is the Honourable Tony Staley, who is also Chairman of the Telecommunications Industry Ombudsman Council.

Council members have been selected for their knowledge of customer interests and customer service issues within the context of the electricity industry.

The main responsibilities of the Council are:

To oversee the EIOV Scheme
To provide advice to the Ombudsman on policy and procedural matters
To maintain the independence of the Ombudsman.

The **Board** is composed of Directors appointed by the member companies and the Board is responsible for the formal operation of the scheme. It exercises final authority for its financial affairs. The Board has no responsibility for the handling of complaints, policy matters, or the day to day administration of the scheme. These matters rest with the Council and the Ombudsman.

The Board is responsible, however, for the formal appointment of the community representatives and the Chairman of the Council but only after wide consultation and in particular with the Victorian Regulator-General.

## 3 The Role of Ombudsman

The role of **Ombudsman** is to receive, investigate and facilitate the resolution of enquires and complaints between customers and the electricity companies. The EIOV can take complaints only after the customer has been to their company and been unable to sort out the problem there.

An extensive range of electricity related matters are covered in the EIOV's jurisdiction:

The provision of, or the failure to provide, electricity (new/existing installations, repairs, delays).
The supply of, or the failure to supply, electricity (outages-spikes/surges, planned and unplanned interruptions).
Electricity bills, the largest area of complaint.
Credit and payment services.
Disconnections.
Refundable advances (security deposits).
Land and land access issues including tree clearing.

Matters raised by the Regulator-General's office.
Matters raised by an electricity company with the consent of the complainant.

The funding basis of the scheme is very much geared towards improving customer service and reducing complaints. Initially, funding was provided by the industry on a market share basis, but from 1 January 1997 funding has been based on the number of enquires and complaints against each company.

The Ombudsman does not deal with complaints regarding:

prices, tariffs or price structures.
commercial activities which are outside the scope of the company's licence.
the content of government policies, legislation, licences or codes.
matters under consideration or previously considered by a court or tribunal.
anything required by legislation, codes, licences or orders.
customer contributions to the cost of capital works.
events beyond the reasonable control of a company, bearing in mind current law and reasonable and relevant industry practice.

The comprehensiveness of the scheme means that it can deal with the relevant conduct of almost anyone connected with the provision of services by a company including its employees, servants, officers, contractors and agents. Not only is the scheme comprehensive, but the Ombudsman has power to make binding decisions where there is no agreement on how to settle a case.

The Ombudsman can make a range of awards in a binding decision, including directing a company to:

pay financial compensation
provide a service
amend or waive a charge
supply goods/services or undertake corrective work
correct, delete from, add to a customer's record
attach a complainant's correction, deletion, addition
do, not do, stop doing an act; or
dismiss the complaint.

Compensation awards of up to $10,000 may be made or up to $50,000 with the consent of all parties.

In looking at the relative state of play of Victorian electricity customer satisfaction, I would like to make a general comment first about how far the industry has come in a short time. When I first took up my position, it was at a time when it was all still a brave new world for Victorian electricity companies, whose policy and practice was guided by what rules, regulations and laws said. Trying to solve a customer issue from a vantage point which focused on the customer rather than on the rule book was for most companies a challenging shift.

Eighteen months later it is pleasing to report how much things have changed. What I see happening in *all* the companies is a serious commitment to trying to solve customer issues at both an individual and a policy level. The various companies achieve this commitment to greater and lesser degrees however. It is a spectrum of achievement and there are opposite ends of the spectrum.

Nearly two years after the opening of the office on 1 May 1996, the industry's maturing understanding of a customer service focus now means that companies are much more willing to explore creative, innovative and flexible solutions to customer problems. Indeed some companies are satisfying to work with as their approach to customer complaints is one of openness and creative solution finding, not defensiveness. They are firmly focused on how to solve customer problems and to learn from their individual cases what the broader issues might be for policy and practice in that company.

At the same time as we develop our mutual relationship of problem solving, complaints still come to my office. In the 1996/ 1997 financial year we received 4,134 enquiries, 550 consultations, 452 complaints, and 30 disputes.

## 4 Ombudsman Activities

### 4.1 Enquires

An **Enquiry** is a request for information or assistance received by phone, fax, personal visit, or in writing. Some Enquires are handled quickly over the phone (e.g. where a customer has not been to their electricity company first to try and resolve their problem, or where the problem is out of jurisdiction of the EIOV). Others may require some further work but can usually be settled within a day and without investigation or contact with the member company. A common example of an enquiry has been explanation of the "supply charge". In some cases an enquiry may be upgraded (escalated) to a Consultation or a Complaint, as appropriate.

### 4.2 Consultation

A **Consultation** is a case that looks likely to settle within the 14 day time frame allocated to this case type, or may be an enquiry which has been escalated. A Consultation:

relates to a member of the scheme
is within jurisdiction of the scheme
has been raised with the relevant member; and/ or
the consumer is dissatisfied with the outcome or the way in which the member has attempted to resolve the issue.

Consultations may require limited desktop follow up such as telephone conversations with the complaints handling staff of the scheme member. Consultations do not involve detailed investigation. An example of a Consultation is a high bill query where the EIOV may request billing details from the company's CIS screens and conduct a phone audit with the customer. Consultations may also be urgent matters which must be settled quickly (eg. imminent disconnections).

### 4.3 Complaints

A **Complaint** may be an unresolved Consultation which has been escalated or may be a case which seems unlikely to settle in the 14 day time frame for a Consultation and requires the 28 day time frame of a Complaint. The conditions applying to a Consultation also apply to a Complaint, however a Complaint is relatively complex and has a greater impact on the complainant than a Consultation. A Complaint will require investigation by both the member and the EIOV and has 28 days in which to resolve. An example here includes a matter in which the customer has a financial claim against the company relating to supply damage or restitution following a tree clearing event or property damage.

### 4.4 Disputes

A **Dispute** is a Complaint in which the relevant member has had sufficient opportunity to resolve and the consumer remains dissatisfied with the resolution of the Complaint or the way in which the member has attempted to resolve the Complaint. Dispute resolution involves a high level of EIOV staff input in active investigation and may incorporate a negotiation/ conciliation role in seeking agreement or a satisfactory resolution on the part of both parties. There is no set time frame for the resolution of Disputes. Disputes which cannot be resolved by negotiation/ conciliation will require a binding decision by the Ombudsman.

A summary of enquires, consultations, complaints and disputes to the Electricity Industry Ombudsman is shown in Figure 1.

### 4.5 Binding Decisions

A **Binding Decision** by the Ombudsman is binding on the company only and not on the customer. The customer has 21 days to accept or reject the Decision of the Ombudsman. If he/ she accepts the Ombudsman's Decision the company must carry out the terms of the Decision. If the customer rejects the Decision the company is released from the terms of the Decision. The flow chart of binding decisions is shown in Figure 2.

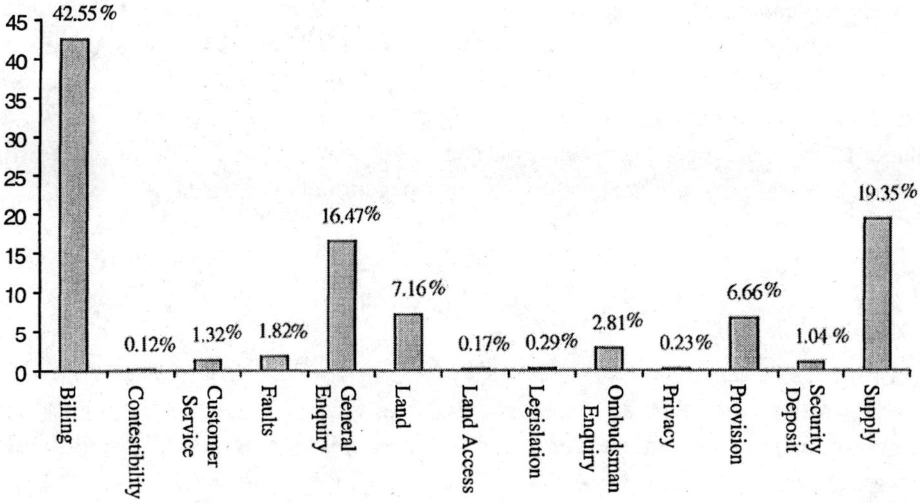


Fig.1 All cases in 1996/97: Enquires, Consultations, Complaints and Disputes.

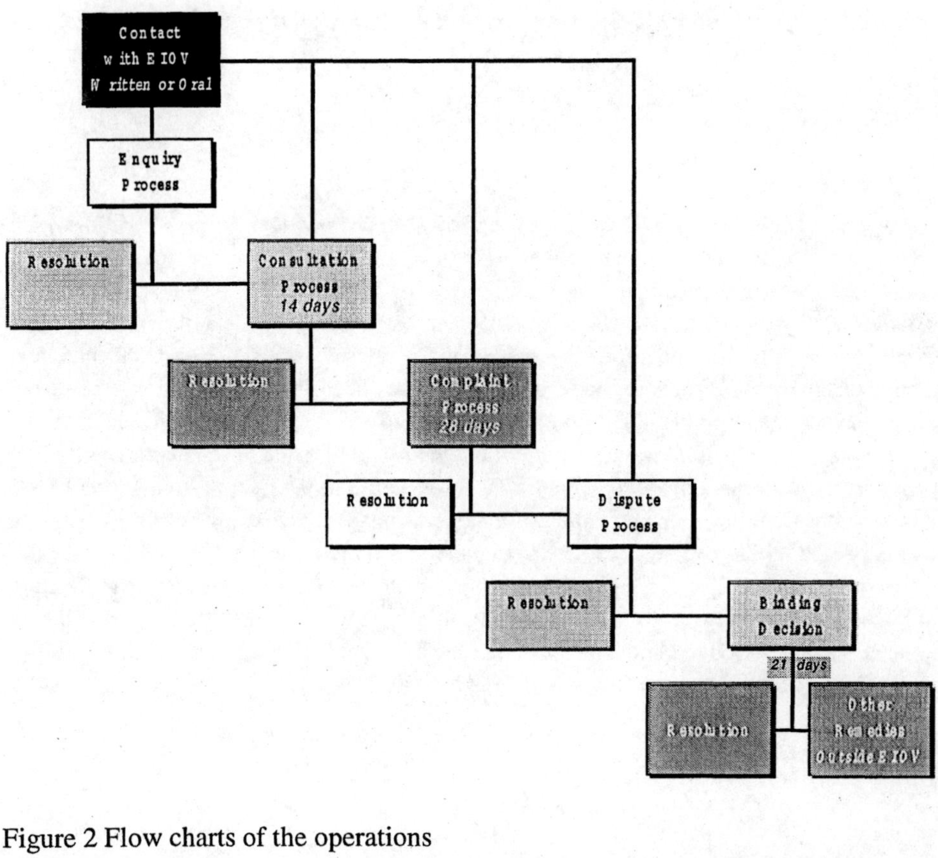


Figure 2 Flow charts of the operations

So what are customers saying about their interface with their companies? It needs to be remembered that my experience is with the minority of complaints which were not able to be solved directly with the company. The vast majority of customer complaints are settled satisfactorily between customer and company. But there is a lot to learn from complaints to my office because that is where things have gone wrong, at least in the sense of there being a dispute, and require some independent help to resolve.

## 5. Solving Problems

### 5.1 Billing

As a proportion of all the case types billing constitutes 42.5% and is the highest area of customer complaint. It's interesting to note that billing cases generally settle within that 14 day category (54%), falling away to 15% in the 28 day category.

Customers raise the following issues: high bills, bills in arrears, backbills, concessions not appearing on bills, non payment of bills leading to disconnection, problems with Easyway timeframes and dollar amounts, billing errors, the format of the bill and whether it is easy to read, lost bills, meter misreadings, estimated bills, reconnection fees charged on bills, explanation of the service charge on the bill, the tariff on which the bill is based.

***CASE STUDY - NUMBER ONE***
***HIGH BILL***

A customer received a bill with an average daily usage of 7.8 kWh per day, totalling $85.90, which she thought was too high. The account related to a new premises and as the customer lived interstate was rarely at the premises during the billing period. The initial account represented a billing period of 60 days, nine days of which the customer stated she had been at the premises. The customer felt the account was not correct because the premises was small and she had few electrical appliances. Her hot water service was gas, and she stated that she turned off the mains switch when leaving the property. The company rechecked the meter, and indicated that the reading was correct, and suggested to the customer that the usage was due to squatters. Unsatisfied with this answer, the customer approached the EIOV. In response to the issues raised by the EIOV as part of its investigation, the company indicated that it was unable to access the customer's meter on the due date of the initial reading. Access was obtained for a meter reading five weeks after the initial reading date was due. This reading was used to calculate an average amount back to the date the initial reading should have taken place, resulting in a high bill. At the time it provided this response, the company offered to charge the customer the average daily usage for the nine days the customer stated she had been at the property, reducing her bill to $45.90.

***CASE STUDY - NUMBER TWO***
***HIGH BILL***

A customer complained that his electricity bill for his holiday house had been too high, given that he had been away for seven weeks. He was dissatisfied that his electricity company had told him that he would have to pay for a meter test if the meter tested correctly.
The company conducted an on-site energy audit. This audit resolved the issue for the customer. The customer was satisfied that the appliances that were operating whilst he was absent from home, probably fully accounted for his electricity consumption and he was happy to pay the account.

So that's what goes wrong with billing, but what about how to get it right? What is it that customers expect?

Companies have a responsibility to provide their customers with better information about the way they bill and the consumption rates of appliances. It is not sufficient to leave the provision of this information to manufacturers. Call centre operators need to be given ample time to satisfactorily explain high bill causes to customers on the phone.
On-site energy audits are an especially cost-effective way of informing and educating customers about their energy usage. It is certainly a cheaper means of addressing an issue than the cost of following through a complaint or having to deal with the EIOV.
As energy reform continues in Victoria with the privatisation of gas, customers should be made aware of the comparable costs of using electricity and gas appliances.
There is ample room for innovation in billing, and some innovative examples exist in Victoria. Some companies issue large print bills for visually impaired customers, and provide bills in a range of community languages.
There is a need each time tariffs change to explain the changes to customers.
As a general rule, customers are not informed about what fees and charges are based on, and this needs rectification.
Many customers still find the bill confusing and would benefit from information about how to read the bill.
Comparative charts and tables are very useful for giving a clear picture of consumption history, and thereby avoiding queries about the higher bill in winter. The charts need to cover at least 12 months history.
Clear information about where and how to pay the bill is important. Many customers still search the bill for this information.
There is room on a bill for special promotions and messages eg. winter bills and available concessions, explanation of particular fees and charges.
Bills create expectations. Changes to a customer's billing cycle will lead to more queries than usual eg. different payment cycles for satisfactory and unsatisfactory histories. Providing information will minimise any confusion experienced due to changes.
Look around and learn from other companies, industries and countries. Find out who is doing billing in an innovative and customer focused way, and learn from their success. One good idea is a flyer to customers that explains their bill.

Estimated bills can result in undercharging, which can set the company and the customer back through the creation of arrears and a loss of credibility in the customer's eyes, as they feel that the company should have got it right in the first place.

### 5.1 Supply

As a proportion of all case types supply cases constitute 19.35% and are the second largest area of complaint. There is a marked increase in the percentage of supply cases as a proportion of the total as you go through the case types. What this says is that supply cases do not readily settle at early stages. Customers raise the following issues with us: outages, either planned or unplanned, and in particular their frequency and duration, high/ low voltage events usually leading to damaged home appliances or equipment, quality and reliability of supply in country areas.

***CASE STUDY - NUMBER THREE***
***UNPLANNED SUPPLY INTERRUPTION TO BUSINESS***

A manager from a processed frozen food business contacted the EIOV regarding compensation for losses including lost products, lost wages and loss of production which he alleged were incurred following an unplanned outage, and lodged a case with a claim for $19,680.19. As part of the EIOV's investigation, the records provided by the company indicated that an unplanned outage had occurred, lasting approximately 11 hours. The company claimed that as this was an unplanned outage, it had not been in a position to arrange an alternative power supply for the customer. Independent technical advice obtained by the EIOV indicated that it appeared that the electricity company's maintenance standards resulted in the ultimate failure of the customer's cable; that lack of an alternative power supply had resulted in loss of stock, and that damage to the customer's computer equipment could have been caused by voltage variations.
Following these initial investigation findings, the company offered the customer an ex gratia payment of $10,000 in full and final settlement of the claim.

***CASE STUDY - NUMBER FOUR***
***FLUCTUATING VOLTAGE***

A customer called the EIOV on behalf of herself and other residents regarding fluctuating voltage in their street, causing for example, appliances such as air conditioners to cut out while in use. The EIOV investigated and the company quickly moved to test the supply quality and following the tests agreed to perform works to shift the supply source to a different substation in order to better balance the load in the area. The company also undertook to assist a major commercial user in the area to upgrade their supply, as this too appeared to be interfering with the customers' supply.

### *CASE STUDY - NUMBER FIVE*
### *FREQUENT UNPLANNED SUPPLY INTERRUPTIONS*

A rural customer contacted the EIOV after being unhappy with the company's response to her complaint relating to recurring interruptions. Outage history details included 30 transient faults involving Auto Reclose Device operation and a traffic accident in the previous 12 months. A resolution was reached with the company making a site visit to the customer to discuss the outage problem and explain the efforts made and being planned to rectify the problem. This included the installation of a Supervisory Control and Data Acquisition System as part of a $5m capital works project by the company. The company appointed a direct contact person to the customer and made an ex gratia payment of $50 on a customer service basis to the customer's account.

### *CASE STUDY - NUMBER SIX*
### *HIGH VOLTAGE DAMAGE*

A customer claimed that a power surge had damaged a number of her electrical appliances with a total damage bill of $2,453.45. The electricity company agreed that the event had occurred, but said that the event was outside of its control claiming the surge had been caused by a possum. The company was unable to provide any evidence of the possum as the cause. The customer's insurance company also refused to pay her claim. The customer contacted the EIOV about the financial loss and inconvenience she had suffered. The EIOV investigated both the technical and the customer service issues in the case. The case did not settle through conciliation and went to a Binding Decision by the Ombudsman. The Ombudsman determined to award $1,741.45 to the customer, taking into account fair repair and replacement costs.

In relations to supply, advices for electricity supply companies resulting from customers' complaints include:

Attend to areas where there is a need for system upgrades. Work with communities to assess the current and future needs of the area. Work with the community to achieve a program of system improvements and negotiate the costing structures for such upgrades directly with the community so that they have a sense of agreement and ownership of decisions.
Explain and promote good work in supply upgrades and improvements. It has been said to me by electricity companies that they don't want to talk publicly about what they spend on maintenance and system improvements as it gives a lead to competitors. The down side of such an approach is that unless you tell customers what good work you are doing to make sure their supply system is safe, reliable and of good quality, they think you are doing nothing.
Promote research and development work. Customers are keen to know what new things are being trialled to give them an even better system. This can be especially helpful where there may be a public issue like power surge damage, where companies can, for

example, publicise trials of new cold transformers which are less inviting for possums to snuggle up to at night.

Pay customers for damage due to high and low voltage variation events. Even though this issue is at different stages of development in the different Australian states and territories, companies are going to have to come to grips with a position that seeks to provide customer service satisfaction, and not be driven solely by what various legislative and regulatory frameworks say.

## 5.1 PROVISION

As a proportion of all case types provision cases constitute 6.7% and this is the third largest area of complaint. The sorts of issues raised with us are: works and connections to existing and new supply sites, generally about delays or dissatisfaction with service.

***CASE STUDY - NUMBER SEVEN***
***REPAIR OF STREET LIGHTS***

A customer called the EIOV because he was having difficulty getting the lights in his street repaired. The customer had a loss of power for four hours one evening and when power was restored he noticed the street lights would not work. The customer called his electricity company several times regarding the lights and was not able to obtain a firm response as to when the lights would be repaired. The company advised that the lights had remained in need of repair for a length of time due to being unable to source necessary parts. The company credited the customer's account with a Guaranteed Service Level amount of $20.

When dealing with the provision of service:

If a company commits to a commencement date for work, stick to it. If it can't, make sure the customer is fully advised of why they can't meet the deadline and what the new deadline is.

- Offer gestures if there is a failure to provide the level of customer service promised. In Victoria there are a number of so called Guaranteed Service Levels (GSLs) which set for the industry five areas which they must achieve, or pay a penalty to the customer in 3 of the 5 areas. If the company is more than 15 minutes late for an appointment the customer's account is credited with $20. If the company fails to meet an agreed supply connection date, $50 per day will be taken off the next bill up to a maximum of $250. If street lighting is not fixed by the agreed date a customer's account is credited by $10. Companies also agree to keep customers informed about supply interruptions, and to provide courteous and timely service. However these last two do not carry any penalties, making customers feel they may be a little hollow. Interestingly two of the Victorian companies have doubled the GSLs as a mark of their commitment to good customer service.

## 5.4 Disconnections

Disconnections account for 1.2% of all cases. We generally manage to resolve disconnections in the early stages, which is not surprising given the need to get supply on as quickly as possible. Most often customers come to us just before they are about to be disconnected because they have been unable to find a payment solution in their direct dealings with their company. Most disconnections are about inability to pay, not unwillingness to pay. Under the Victorian Supply and Sale Code, once a complainant has lodged a complaint with us a company cannot proceed to disconnect.

We also get customers ringing who have been disconnected. We try to get them back on supply and resolve the outstanding issues, without the background of no power supply to taint the resolution process.

### *CASE STUDY - NUMBER EIGHT*

***DISCONNECTION***

A customer claimed that he had been disconnected unfairly, that he had recently been declared bankrupt and that the company was one of his creditors. It had asked the customer to pay a refundable advance of $200. Six weeks prior to the disconnection, Social Security had negotiated on the customer's behalf and the company had agreed to waive the refundable advance. The customer was paying his accounts via an Easyway plan, in fortnightly payments of $20. When he subsequently received a bill for $1.38, he discontinued his payments as he believed he was now in credit. The customer was then disconnected for an unpaid amount of $201.38, $200 of which was the refundable advance. When the customer called the company, he was told that supply could only be restored if he made a payment of $144 that day ($44 reconnection fee, $100 refundable advance). The customer felt that this was unfair as the company had previously agreed to waive the refundable advance and had billed him for only $1.38. The EIOV, as part of its investigation, examined CIS records and requested other relevant notes from the company. The company provided the requested documentation but moved to quickly offer a resolution. The company agreed to waive the refundable advance, reconnect the customer that day and not charge a reconnection fee. The customer agreed to bring his account up to date within the next week and to adhere to the plan in future.

There should be greater focus on how to prevent disconnections. They're not cost effective for companies or the customer, and result in a whole set of processes and fees that can be avoided if performance in credit management is improved. Victorian companies have had considerable success in introducing even the simple new step of ringing customers in arrears and reminding them personally that their bill is due or that disconnection is imminent. This has helped contribute to a reduction in disconnections and improved cash flow.

## 5.5 DISCONNECTIONS

As a proportion of all case types refundable advance cases constitute 1%. This is a category where small business customers are the main callers. They express concern about how the amount of the advance has been calculated, the period for which the advance is kept, and previous successful business histories not being taken into account. An electricity retailer should:

Explain how the advance has been calculated and how the period for which the advance is levied is worked out. Do this when the advance is being required and when it is being returned. We have had cases where the notification of the return of the advance was so small on the bill, the customer did not know it had ever been returned.
Be flexible in assessing previous successful business history and work with each customer on a case by case basis.

## 5.6 CREDITS AND PAYMENT SERVICES

Credit and payment services are included in the billing data and account for 1.6% of all cases. Generally cases cover disagreements about how to structure payment arrangements, the dollar amounts and the payment schedule. Occasionally there have been problems with services provided by contractors who handle credit or payment arrangements for companies.

***CASE STUDY - NUMBER NINE***
***BILL PAYMENTS***

A customer was told by her electricity company that a payment of just over $100 she had made through the company's payment contractor had not been received. Investigation revealed that the payment contractor had not recorded the payments correctly, including transferring some of the payments to the customer's landlord's electricity account by mistake. This case was settled by the company waiving the $100 debt.

The conclusions from dealing with such problems are:

The old ways of credit management are not going to serve companies well into the future. The industry needs to be creative and innovative in the way they address the prevention of arrears, instead of addressing arrears once they have been allowed to happen. Find ways to work with customers to help them manage their electricity accounts. Many customers find large quarterly bills hard to pay and both the company and the customer can benefit by introducing payment options such as instalment plans.
Make links with relevant community agencies like financial counsellors. Some companies in Victoria have taken the innovative approach of funding projects to identify

impediments to ability to pay in their local communities. They work directly with community agencies to assist customers in the management of their finances, thereby preventing situations of credit conflict between customer and company.

For customers with severe payment difficulties it may be helpful to try creative solutions such as incentive plans where the customer is rewarded for keeping to their payment arrangement.

There is no point in seeking a payment arrangement without regard to the customer's ability to pay and without reference to an independent financial adviser such as a financial counsellor to properly assess the customer's capacity to pay.

## 5.7 Land and Land Services

As a proportion of all case types land cases constitute 7.33%. Here we are looking at easements, pole maintenance, street lighting, and tree clearing.

Ensure your customers know about your programme of maintenance for all your assets: the poles and wires, and street lighting.

Consider street lighting gestures such as the Victorian GSLs discussed previously.

## 5.8 Tree Clearing

Tree clearing cases are the major component of the land and land access cases. We get cases from both urban and country areas about unauthorised tree clearing/ trimming, lack of notification of tree clearing activities, damage to property during tree clearing activities, and rude/ abusive behaviour of tree clearing contractors.

***CASE STUDY - NUMBER TEN***
***TREE CLEARING***

A customer contacted the EIOV because he was unhappy with tree clearing that had taken place on his property and the power company's response when the issue was raised. The power company's contractors visited the customer's property unannounced and trimmed two established trees, a 20 year old Silver Birch and a 10 year old Eucalypt. Three months previously, these trees had been trimmed. At this time the customer was told by the company that the trees would not have to be trimmed for another two years. At the time of the tree clearing, the customer's wife came out to find out why the trees were again being trimmed. Her queries were ignored.

She then contacted the company and asked to speak to a supervisor. The customer was dissatisfied with both the supervisor's response and attitude, which had left her feeling distressed.

The customer contacted the EIOV requesting that the issues be investigated. The investigation revealed that the initial tree clearing at the customer's premises in October 1996 had, in error, not been recorded. The customer's property was on the market and the actions of the company in clearing the trees had detracted from the property's value. The company offered the customer a

$1,000 no liability payment. The company also extended its apologies to both the customer and his wife.

To reduce the number of complaints, a supply company should:

Ensure notification procedures are in order to ensure that tree clearing/ trimming activities do not create customer anger due to lack of consultation and advice.
Consider providing easy to read information about the relevant tree clearing codes and regulations so that customers can see where you are coming from. This information should include clarification of who is responsible for which areas, such as local Councils.
Work with local communities on tree clearing activities.
If this function is contracted out, ensure contractors are properly briefed about how to operate in the field and in proper customer relations.

That about covers the how to's, but while we are on the subject of contractors I would like to briefly give you some insights into how to avoid contractor/ agency problems which have come to our notice through complaints to my office:

Customers are not interested in a shifting of responsibility for mistakes and problems from the company to a contractor. They rightly hold the company responsible.
A company's contractor or agent is their image in the market.
For contracting and agency agreements to work effectively, time needs to be spent in teaching contractors what is expected of them to perform their tasks in the way you want them to be performed, particularly policies and procedures for: notification of events; means of communication or consultation with customers; what to do when a customer is unsatisfied with a particular service/process; how contractors are to behave in the field.
Both parties to a contract or agency agreement need to learn about each other's policies and procedures and ensure that the policy is carried out in practice.
One of the critical issues in entering a contractual arrangement is to decide what if any direct contact contractors may have with customers. If direct contact with customers is likely (eg. tree trimming), then time and energy must go into ensuring that contractors behave in a way that is expected of company staff.
Excellent communication channels need to be established between companies and their agents, so that proper notification and agreed prior arrangements are communicated to contractors and agents doing work on behalf of a company.
If a contractor or agent is not performing, then change contractors/agents.
There is a need to establish methods of evaluating the performance of the contractor against performance standards. Customer service should be one of these standards.
Consideration should be given to whether there should be an incentive for good performance or a disincentive for bad performance.
Does the tender brief ask for potential contractors/agents to make clear in their proposal what their policies and procedures are for providing that service?

The last issue to think about is internal complaints processes. They can be a great source of knowledge and information about how to improve service. Do customers know what the internal process is? Are they advised to pursue it in the event of a complaint? Does the organisation think of complaints as a burden or a valuable resource from which to learn more about company procedures and policies?

## 6 CONCLUSIONS

Victorian small business and domestic customers will be able to choose their supplier on 1 January 2001, and let there be no doubt that customers in the last two tranches of the contestable market (1 July 1998 and 1 January 2001) are looking with great interest at what has been happening in the other tranches. They are quietly (and in some areas not so quietly) filing away information mentally on which company has been reported as doing what lately, how it has performed in the Office of the Regulator General's performance reports, whether it has many outages and surges, and how customer friendly the company is when you have a problem. So even though 2001 seems a long way away, domestic customers will come to that day with knowledge of who they think has performed well and a sense of whether they want to change retailers or not.

I hope that each of you will ultimately be able to tap into the value of independent data about customer service performance in your company through some kind of independent alternative dispute resolution mechanism, from which you can learn much about how to satisfy your customers and their issues, and at the same time improve your competitive advantage.

****** 7 ******

# ENERGY USERS' EXPERIENCE

**ROMAN DOMANSKI**
Executive Director, Energy Users Group, Australia, Email: Doma@wizard.teksupport.net.au

**Summary:** This paper outlines the activities of the Energy Users Group and its experience with competitive electricity markets. The Group was established as a non-profitable organization to represent interests of energy users on a range of policy, regulatory and energy supply industry issues. It also acts as a forum for discussion and information exchange among end-users in Australia.

## 1 INTRODUCTION

The electricity supply industry has undergone fundamental change over the 1990's. The industry which started the decade almost totally in government ownership and as a series of State monopolies, has been transformed into a competitive market, where private ownership is becoming more common.

## 2 CUSTOMER CHALLENGES

For major electricity users these dramatic changes have also thrown-up new challenges. Whereas previously there was little else to do but wait for your "friendly" public utility to come along and tell you how much you were going to pay for your power, now there is the complication of having to negotiate a contract for electricity with a Distribution Business (DB). There is a difficult choice of whether to "go with" the local DB or switch to another one, the complexity of having to wade through thick contract documents, the decision about whether to contract for supply in-house or whether to use one of the many new service providers and re-sellers who have emerged. It is also possible for end-users to join in the electricity spot markets, which have been created in Victoria and NSW (until now, only one end-user has joined the market). Eventually, smaller business customers, from July 1998, and even households, from year 2000, will also be exposed to the competitive market.

These challenges are daunting for energy-users and require a range of new skills, resources and operating methods. In open electricity markets new skills such as risk management of electricity, managing ones demand profile better and more efficiently, and skills in negotiation of hedge contracts are needed. Many companies are quite used to negotiating contracts in other areas of business and can apply these skills to an extent, although electricity is somewhat different.

## 3 THE ENERGY USERS GROUP

The Energy Users Group (EUG) is an association of energy users formed to represent the interests of its members on a range of policy, regulatory and industry issues. It also acts as a forum for discussion and information exchange among end-users.

The EUG is a non-profit organisation funded mainly by membership fees and focused on energy issues. The members determine EUG policy and directions. The Group is a national body of business users and associations with members and activities across all states in Australia (see Appendix A.1).

### 3.1 MISSION AND OBJECTIVES

The EUG's mission is to become the voice of energy users in Australia and to effectively represent the views and concerns of its members. The key objectives of the Energy Users Group are to:

- Improve outcomes for its members from energy reform.
- Ensure that these reforms deliver sustained lower prices, customer-driven service and continued high reliability through a competitive and efficient energy market.
- Ensure member's views are well represented to regulators and policy makers.
- Provide a range of relevant services to members in order to assist them to operate more effectively in the reformed energy market.

The Group places a high priority on ensuring that its work is well introduced and credible. Work priorities are set out in Appendix A.2.

## 3.2 HISTORY

The EUG was formed in 1996 by the Australian Business Chamber, Australian Chamber of Commerce & Industry, Australian Chamber of Manufactures, Business Council of Australia, and Minerals Council of Australia. Since then, it has steadily expanded its membership base to the point where its members are mostly direct energy consumers and it is driven by their views. It was formed to provide a focal point for developing end-user views on energy market reform and in recognition of the increasingly detailed level of issues under debate.

## 3.3 SERVICES

The EUG provides members with a range of services/advantages including:

networking and information exchange among end-users;
regular meetings to discuss customer-based issues;
representation to regulators and policy-makers;
representation to energy companies on customer-wide issues;
regular updates of developments in the energy market; and
seminars and workshops on customer issues.

## 3.4 SHARING INFORMATION

The EUG provides members with an important avenue to network with their peers in other companies, with the energy market more broadly, as well as with regulators and policy-makers. It does this through regular meetings and other events, as well as by improving the lines of communication between end-users. Members can exchange information for mutual advantage, discuss issues, learn about the experience of others and improve their ability to operate effectively as energy buyers, managers, or as people that interact with them.

For example, the members have been able to put a common position and influence debate in NSW on network connection, emphasising the importance of 'plain English' contracts, equal allocation of risks between purchaser and provider, best practice performance standards and performance monitoring. The EUG is involved in similar issues in Victoria through the review of the Distribution Code (the document that regulates quality of supply) and expects to be similarly involved in Queensland and South Australia as these markets are opened to retail competition from 1998.

### 3.5 Representation to Regulators and Policy-Makers

The Group has considerable expertise in representing the interests of members to policy-makers and, the increasingly important regulators. Over the coming year, important outcomes can be delivered to members in this area, with issues such as the Australian Competition and Consumer Commission (ACCC) decision on the National Electricity Code. The full implementation of the national market, the National Electricity Code Administrator (NECA) network pricing review, Office of the Regulator-General (ORG) pricing review, and state reforms in Queensland, South Australia (SA) and Western Australia (WA) also require attention.

The campaign against the New South Wales (NSW) electricity levy, which was introduces in May 1997 sups $100m annually from contestable customers, and similar imposts in other states will also continue. This could require additional external funding, an examination of legal issues, and more joint efforts with other interested groups

### 3.6 Dealings with Energy Supply Companies

During the 1997, the EUG became more involved with the electricity utilities through representing the interests of users on customer wide issues such as network connection, network augmentation, metering, setting up processes for the national market, dispute resolution mechanisms, futures contracts, and retail settlements.

The coming year will see the Group continue its involvement in such matters as reforms are implemented and start to shape the market. For example, the connection agreements issues will need to be completed, metering will become more important as retail competition is expanded and customer aggregation issues will become more critical as jurisdictions increase retail competition. These issues have been built into the EUG's work program for 1997/98.

### 3.7 Regular Updates of Energy Market Developments

A variety of services are offered to keep members informed of the latest developments. For instance, the members are provided with information on market issues such as pool prices, the emerging national market, and state regulatory issues. Minutes of meetings and reports on other key events are circulated so that all members are kept aware of issues and discussion. Over the coming year, it is planned to enhance these services. Seminars and workshops are planned to provide additional information to members. Additonal market education and training programs to help members better understand the operation of the reformed energy sectors will also be considered.

## 3.8 Seminars and Workshops

The Group and its members have carried out and participated in workshops and seminars on a number of end-user issues. During 1997/98, this service will be extended to cover a range of customer issues such as the ACCC's authorisation of the national electricity market, the role of the National Competition Council in electricity reform, network pricing, metering, reform in South Australia and Queensland, electricity price trends, service quality/reliability, demand management, contracting, hedging, and environmental issues. A major customer-focused conference is also planned.

# 4 Service Features and Comparison

The main services provided by the EUG and their further extension and development were described above. The Group already provides a range of services to its members and further developments over 1997/98 are planned.

The Group has the ability to provide a range of services not offered elsewhere. In particular it:

has direct access to members who are actively involved in the energy market as buyers and understand the issues that are important to customers;
can provide analysis and assessment of the policy and regulatory issues down to a level of detail and complexity not easily reproduced by other organisations;
can make a stronger impact on decision-makers and the market place by allowing customers to act jointly on a range of issues;
has considerable expertise in the detail of electricity market reform, which is not easily replicated by other user bodies (including in the increasingly important national market) and directly relevant to other areas (eg State reforms).

# 5 Benefits of Reform

Electricity users have benefited through having access to competing sources of energy. For example, the Australian Chamber of Manufactures (who are members of the EUG) recently surveyed customers in Victoria who have been exposed to the so-called "contestable market" - basically where they get a choice of a supplier. It found that 78% of these customers had experienced price reductions, with the average reduction being 10%. This is consistent with the information available to the EUG and larger reductions are generally available to consumers prepared to exercise their competitive right and switch retailer. Price reductions up to 20% - 30% are not uncommon among energy users. In addition, the Electricity Supply Association of Australia found in the more recent survey that electricity users in NSW and Victoria that are able to chose their retailers have had average energy price reduction of 35%. The EUG is therefore a strong supporter of energy market reforms.

Nevertheless, users are concerned that these reductions should be sustained in future. Some benefits may not last because there is still excess capacity in generation and because retailers have been prepared to offer large reductions in the first round to secure a larger customer base. It appears that about one-third of customers eligible to do so have changed retailer. International benchmarking also shows that Australia is still below international best practice in electricity prices, particularly when allowance is made for our abundant energy resources. For example, in the first quarter of 1995 we ranked sixth among OECD countries for industrial tariffs and fourth for domestic. This is not bad, but it could be better!

There is also some evidence indicating that customer price benefits have been more limited when account is taken of the available gains in efficiency from reform.

In addition, network charges remain too high and are not well structured. The quality of network services is also too variable and not always up to customer expectations.

Users are also concerned that electricity contracts are very complex and distribution companies, who are also licensed retailers, are using their still considerable market power in setting contract terms. This is pushing unjustified market risk onto customers. The utilities also do not offer sufficiently competitive contracts to some types of customers, in particular, customers with many small sites which are aggregated into a large overall purchase.

## 6 Attitude to Privatisation

The EUG strongly supports the introduction of competition into the power business. Users believe that this is the best way to ensure that we have continued access to competitive power in Australia - and given our abundance of energy resources there is every reason why we should have. Privatisation is, in fact, an essential part of getting competition, because without private ownership there can be no genuine competition - the assets would all be owned by the one shareholder.

However, there are also risks to competition in privatisation. One major risk is that, in selling electricity assets, a government will be very tempted to increase the price it gets and will compromise on competition to do so. Another risk is that privatisation makes future changes to enhance competition more difficult, as is being shown in the UK and Victoria, because the asset owners object to changes which expose them to more competition. If government does compromise on competition, electricity consumers will be the ones left "carrying the can", which is exactly the opposite of what successful reform of important infrastructure like power is all about.

Victoria has introduced the most far-reaching electricity reforms in Australia. Measured in terms of the sale prices, the Victorian electricity privatisation have been very successful, having realised significantly more than was expected. However, if as seems possible, some competition has been sacrificed along the way, then the impact may only become obvious sometime after the event.

As a group, we are concerned that the conditions under which assets are privatised do not start to detract from competition. For example, co-generation projects already have difficulties in negotiating network access charges and many of exemptions from the national electricity market will diminish competition.

These challenges will increase as NSW, SA and eventually other states begin to sell their electricity assets. The challenges are particularly important for users because the market structures are not yet competitive enough.

## 7 TOWARDS THE NATIONAL ELECTRICITY MARKET

The National Electricity Market, which started in May 1997, joins the power systems of NSW, Victoria and SA and eventually Queensland. This initiative has the potential to move Australia out of the straight-jacket of state power markets and into a larger and much more competitive national market. As such, the EUG is a fierce advocate of a genuinely competitive and genuinely national market. However, the word to emphasise is genuine, and a competitive national power market should not become a scapegoat for private ownership.

For the market to be a "genuine" national and competitive one, there will need to be close attention paid to the following:

the Code of Conduct, which will determine how well the market operates needs to undergo further important changes and must be allowed to evolve over time - the current version is still too complex, will inflate transmission and distribution charges, does not allow transmission and distribution charges to reflect the true cost of using the system, imposes costs and other barriers to competition and in many ways resembles more a series of State markets rather than a national one;

the ACCC (Australian Competition & Consumer Commission) had a very important task in assessing the Code as an 'independent umpire' and some good improvements including a right to by-pass existing networks, due to input from end-users and new entries.

competition in power generation is important (as the lack of it in the UK shows) and should be advanced further - in particular, there is a real threat of domination of the national market by the three large generators in NSW (all owned by the one shareholder - the government) which is in need of further attention; generators in SA and Queensland still have too much market power

also, co-generation can be a useful source of additional competition - and is environmentally friendly - but fails to get a fair go in the national market; and

retail competition is directly important to energy users, but is excluded from the national market - one simple but important step is that the States should recognise each others licensed retailers, so that if you have a retail licence in one State it is good in another; thereby encouraging larger scale and more competitive national retailers.

The EUG is working constructively with Governments and bodies such as the ACCC, State regulators and the two new bodies created as part of the national electricity market (NECA and NEMMCO) with the aim of further reforming the electricity industry and, most importantly, shaping it into a competitive market.

## 8 Conclusions

The EUG has a growing membership made-up of energy-users, business associations and others who share these objectives. We seek to further the interests of energy-users and newcomers - such as co-generators - who offer greater competition, particularly by ensuring that the benefits of electricity and gas reform flow through to users. The Group has a vital role to play in the deregulating energy markets now being created and in shaping their future.

The EUG finished 1996/97 with 16 new members having started the year with 6 members. It currently has 23 members. Further expansion in membership during 1997/98 is very important to ensure that the Group remains effective in the important issues that face energy customers. However, to achieve expansion the Group must put resources into areas where it can make a difference to customers.

There is great potential for the EUG to grow over the next few years. Expansion is certainly possible given that there are currently some 6,000 contestable customers (>750 MWh per annum) in NSW and Victoria. From next year, a further 16,000 become contestable in these two States (>160 MWh per annum) and contestability starts in Queensland, SA and the Australian Commonwealth Territory. Certainly, there is great potential for the EUG to be a strong and effective voice for energy users.

## Appendices

### A.1 Members of the Energy Users Group

Air Liquide Australia Ltd
Amcor Ltd
Ampol
Australian Business Chamber
Australian Chamber of Manufactures
Australian Cogeneration Association
Boral Ltd
Brisbane City Council

Business Council of Australia
Department of Defence
General Motors Holden
Harvey Norman Holdings Ltd
ICI Australia Ltd
Kemcor Australia Ltd
Melbourne Water
Minerals Council of Australia
Newcrest Mining
SA Water
Shell Australia Ltd
Sydney Water
Telstra
Visy Paper Ltd

## A.2 OUTLINE OF MAJOR ACTIVITIES FOR 1997/98

| TOPIC | SUB-TOPICS | PLANNED EVENTS | INDICATIVE TIMETABLE |
|---|---|---|---|
| DEMAND-SIDE PARTICIPATION | wholesale market<br>network pricing<br>role of retailers<br>aggregation | Survey<br>Forum<br>(possible follow-up topic workshops) | Feb-Mar '98<br>Feb-Mar '98 |
| ACCC CODE APPROVAL | draft decisions<br>final approvals | Briefings | Sept 1997<br>Dec 1997/Feb 1998 |
| STATE REFORMS | retail market<br>aggregation<br>Qld market<br>SA market<br>role of NCC | Seminars | Mar-Apr '98<br>Apr-May '98<br>Feb 1998<br>1st qtr '98<br>2nd qtr '98 |
| CONNECTION AGREEMENTS | Quality & reliability<br>rights & responsibilities<br>best practice<br>state issues | Seminars | Before end of '97 & 1st half '98 |
| METERING | Costs<br>Standards<br>Communications<br>Providers | Seminar | 1st half '98 |
| NETWORK PRICING | transmission<br>distribution<br>pricing methods<br>derogations<br>by-pass<br>cost reflectivity | Seminar/conference | 1st & 2nd qtrs '98 |
| CUSTOMER CONFERENCE | Major event | Annual conference | 2nd half '98 |

# Part 3 Trading Electrical Energy

****** 8 ******

# GENERATOR EXPERIENCE IN OPEN ELECTRICITY MARKET

**CHRYS CHANDRARAJ**
Market Analyst, Southern Hydro Ltd, Melbourne, Australia, Email: Chrysc@melbpc.org.au

**Summary:** The paper presents the experience of Southern Hydro Operations. a company trading electricity, in a competitive open electricity market. The operation of the hydro electric plant of the company can be considered by the various types of generation modes outlined in this paper. This relates to how the plant operates, and what are its limitations. The advantages and deficiencies of the hydro plants in competitive markets are considered. In conjunction, the risk management strategies, and advantages of energy contracts to manage market risk are discussed.
This paper considers the following:

- Different types of schemes of Southern Hydro Operations.
- Physical plant operation and limitation.
- Open electricity market experiences.
- Advantages and deficiencies of the different schemes and strategies.
- Market risk issues faced by the company.

## 1 INTRODUCTION

The main issues faced by a hydro electric peaking generator in an open electricity market include:

- Water Management
- Pool Trading
- Risk Management

The performance in the electricity market depends on operation during high demand periods in the overall year. How does hydro return a higher rate for energy usage is considered along with how does peaking plant compare to base load plant. (The water management issues were there even in times before the open electricity market).

Market Risks faced by Generation can be considered in the categories below:

- Uncertain Generation
  - ♦ Forced out of Market by lower bids
  - ♦ Water or Plant Availability
- Uncertain Pool Price

To reduce market risks, market participants can use various financial instruments. Typically, these instruments are contracts between generators and customers. The contracts are traded as "Over the Counter" (OTC) contracts. They usually specify the following quantities:

- Set power quantity (MW) over trading periods
- Variable Energy Price
- Option Fees

The most common contracts are:

- Two Way Contracts. The contracts operate with a strike price for a set amount of MW capacity. The contracts are usually firm, and amount to a straight financial instrument.
- One Way Cap Contracts. These contracts have a strike price when they become effective. The preset MW capacity in effect has a limited exposure, when the price exceeds the strike price difference, payments are made to the counter party.
- Collar Contracts. These contracts operate between two strike prices.

Electrical energy contracts can be arranged in two characteristic ways:

- Firm contracts. Firm contracts operate irrespective of generation.
- Non Firm Contracts. Non firm contracts only operate when the contracted quantity actually generates.

Participants may have a portfolio of contracts with each other and/or a third party. Third parties may take up any mismatch between participants. Examples of third party intermediaries include:

- Morgan & Stanley

- Macquarie Bank
- WestPac

## 2. Physical Operation of Hydro Scheme

Southern Hydro generation, located in North-Eastern region of Victoria is a collection of energy limited power stations. Its generation amounts to approximately two percent of Victoria's energy of about 950 GWh calculated as long term average generation, see Fig. 1.

The objective of water management and energy trading is to optimize a limited resource in order to maximise the revenue. This strategy may be different to plant with unlimited energy, particularly with respect to contracting of energy.
Hydro generation is generally very flexible. There is no long start up process involved. The quick responsiveness of hydro generation expects to earn premium returns. The fuel source for hydro generation is weather dependent (rain/snow/irrigation). The irrigation schemes run opposite to hydro electrical schemes. When there is a lot of rain the irrigation is utilised less, while the hydro electrical schemes will generate more. The situation is reversed during dry weather periods.

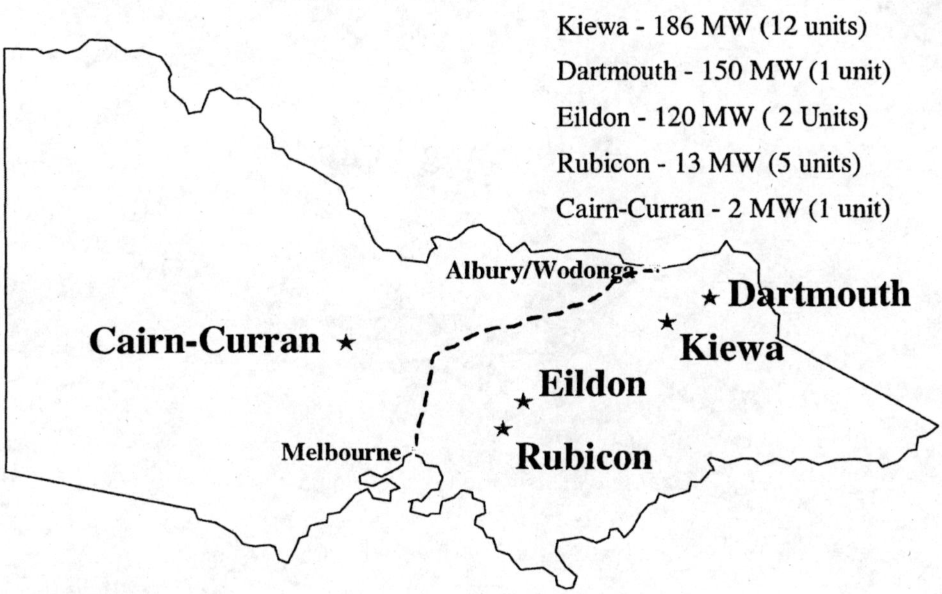


Fig. 1 Location of Southern Hydro Generation sites

The irrigation schemes have a feature that allows extra generation using entitlement water. This feature may be used when the electricity system is short of generation capacity, and values generation very highly. There are synergies in the portfolio of generation that Southern Hydro operates, and it uses this to maximise its revenue in the market. The flexibility of operation does have constraints. These are irrigation and hydrological in the main.

The irrigation authorities control water flow as a result of demand for water by farms located downstream. This relates mainly to summer demand and periods of very dry weather. Flexibility available through use of regulating pond and entitlement water enables the company to meet its irrigation requirements whilst still managing its electricity generation and risk management functions.

The schemes have large storage reservoir. Irrigation authorities determine (river) release requirements from the regulating pondage to maintain a required irrigation supply/river flow. Flexibility of operation is dependent upon the ability to utilise the regulating pondage to maintain irrigation releases whilst varying generation output, as seen in Fig. 2. As pond level falls, the generator is run, releasing water from storage into regulating pond to maintain constant flow of river.

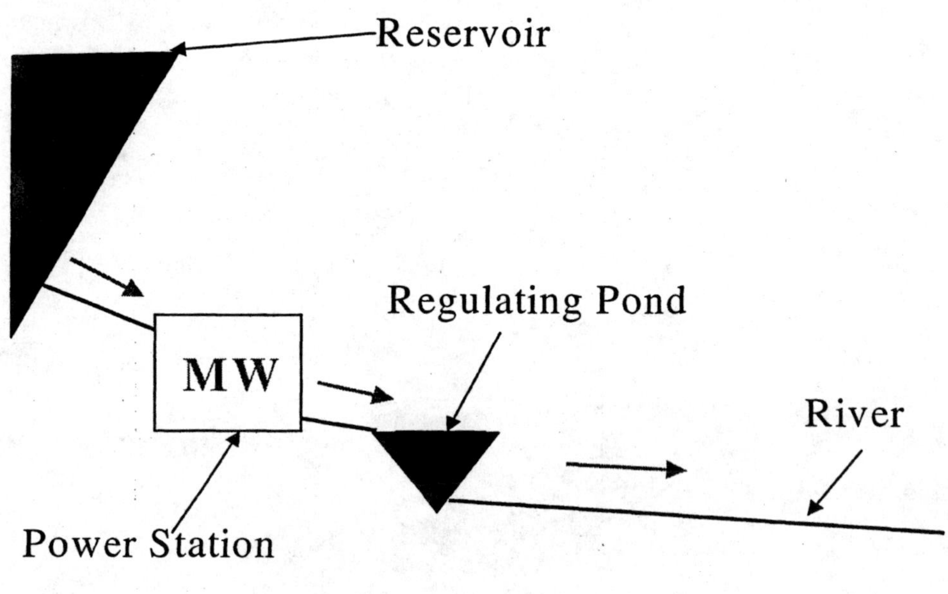


Fig. 2 Example of Irrigation hydro scheme

The generator can run continuously at a low level of output to maintain the river level. Else it can be run at a high level for a short time, and switched off to allow the regulating pond to reduce to a level to allow the plant to generate again. This gives the flexibility to alter the generation pattern to respond to price changes in the market and enables the generation to be flexible while the river flow is constant. Some use of discretionary water, which Southern Hydro can call above irrigation requirements, is also available to meet market changes.

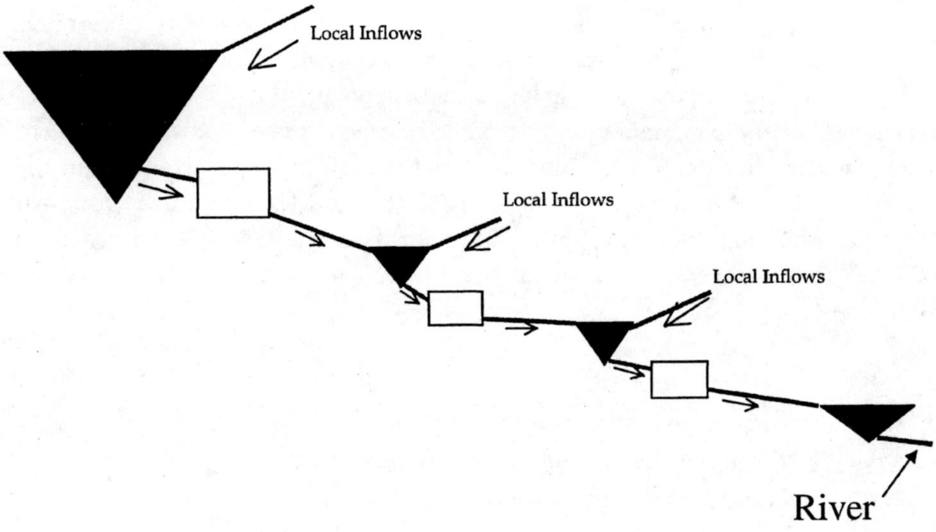


Fig. 3 An example of cascaded hydro electric scheme

In the cascaded hydro scheme water run in the top station flows through all stations located below. The main storage is located at the upper dam, while smaller storage is constructed down the scheme. The flows below the main storage effectively needs to be run within the day or few days. Water into the main storage can be stored for a longer period. Generation from the top generator requires the lower generators to operate. The regulating pond at the bottom allows the river flow to remain reasonably constant, as seen in Fig. 3.

The hydro generation can be considered as "Run of River". This scheme is employed when it rains utilising water which "flows naturally" in the river systems. It provides higher outputs in wetter months with effectively little to no storage capacity hence scheme can works when the river runs.

In summary, the plant availability is dependent on the weather. Hydro generation is very flexible and quick responding, with a fast start capability and is very reliable. There is minimal loss of energy when plant does not generate.

## 3 Market Generation

A bidding strategy is to optimise hydro generation over long and short term. The objective is always to generate when prices are the highest. It is very important to determine trading periods when generation should take place to achieve the objectives. Other factors like trade off volumes needs to be considered when determining the bidding and contact strategies.

Hydro generation do not generate to meet contracts, but have available the capacity to meet contracts at times when the pool price is greater than the strike price. The generation does not aim to set price, but maximise return. This will take place when hydro generation is marginal and can influence the pool price.

There is a very important role for forecasting pool price over the long term time horizon for planning purposes. This will assist in making water available in the best times to maximise revenue. There is also a need for short term pool price forecasting in order to maximise the generation in the short term. This is usually in the day or a week.

There are three characteristic generation modes considered by Southern Hydro (see Fig. 4).

- ***Forced - Base load generation.***
- ***Semi-Forced Generation - Intermediate/Peak.***
- ***Unforced Generation – Peak.***

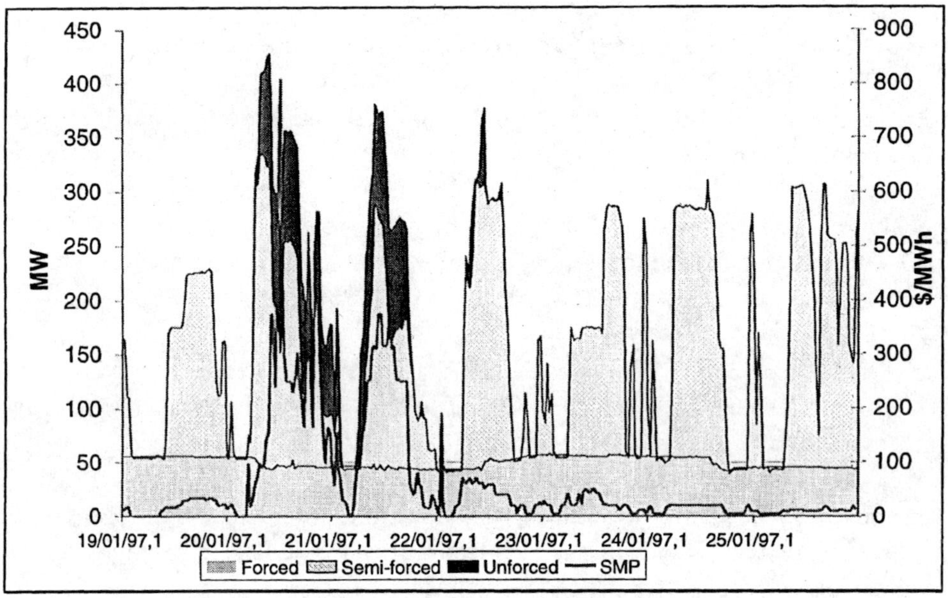


Fig. 4 Example of Hydro generation is a volatile pool price week

| | | | |
|---|---|---|---|
| Average SMP | = $62 | Forced rate | = $58 |
| Average Peak SMP | = $87 | Semi-forced | = $83 |
| | | Unforced | = $217 |

## 4 Responding to Market Price Deviation

Aiming at the maximum return a price optimization strategy is undertaken. This depends on a variety of factors:

- Rate varies with System Marginal Price (SMP) level. If SMPs are high then you make earn a higher rate, when SMPs are low you earn a lower rate. This is a function of the SMP level.
- Rate varies with SMP volatility. If SMPs were, all the same level then you can only earn that level. This is a function of the volatility of pool price. If SMPs vary a lot then you can earn a higher rate of return if you can concentrate the energy into the higher price times.
- Rate varies with energy type. Energy that is required to run for significant periods (base load) will only earn the "average" rate. Energy that only runs for short periods has the potential to earn a higher rate.
- Rate varies with responsiveness.

An example of such a strategy is MKPS generation, which is part of the Kiewa scheme, generating at higher prices of a month. On the other hand, the irrigation schemes such as Dartmouth and Eildon spread over month due to irrigation requirements. This has resulted in an annual premium to the SMP of 43% for the 1996/97 fiscal year.

The risk management objective is to regulate a portion of the revenue stream of the company. A proportion of the hydro generation needs to be hedged in forward contacts to earn a steady revenue and not be completely exposed to the spot market. There is the risk that pool prices would not ensure sufficient return to provide a viable business. The risk issues are as follows.

1. Risk of unfavorable Pool Price
2. Risk of Hydro Generation
3. Risk of Contract Exposure

The risk management strategy of the company considers these issues, in developing policies to mitigate the risks. Strategies implemented can be summarised as follows:

1. Contract Hedging
2. Contract Exposure

The main objective of hedge contracting is to increase cover for forced generation and obtain favourable contract prices. As the contract cover increases to include more flexible water flow, there will be an expectation of higher contract premium. Cover the energy exposure of contracts by having energy available (if need when pool prices exceed the strike price), but not necessarily generating. Closing out contracted exposure (of energy and capacity) if lack of generation.

The Southern Hydro strategic advantage lies in its flexibility and responsiveness of generation to attract premium returns for its generation.

Contracting strategies depend on the type of generation that is being contracted, and what the generation would earn on the pool. The types of generation are as follows:

1. **Forced – Base.**
2. **Semi-Forced - Intermediate/Peak.**
3. **Unforced – Peak.**

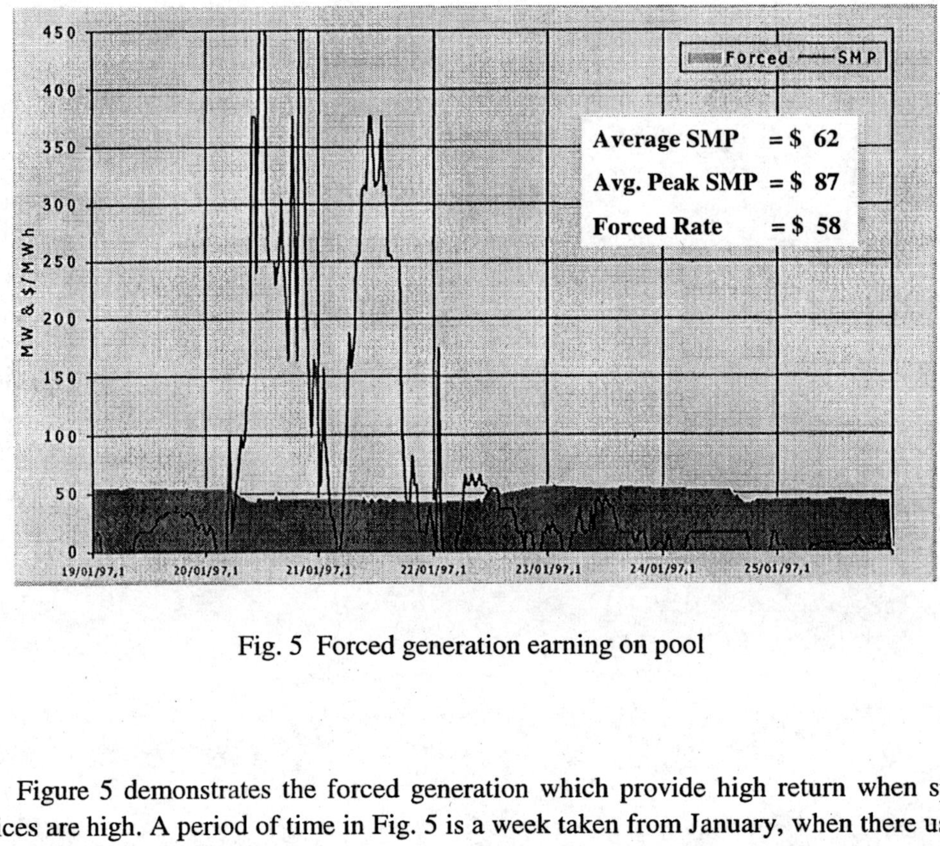


Fig. 5 Forced generation earning on pool

Figure 5 demonstrates the forced generation which provide high return when spot prices are high. A period of time in Fig. 5 is a week taken from January, when there used to be volatility and high prices in the pool. Forced energy generation is effectively generation that has to be run "all day". There is no special response to the higher prices or shifting of energy as the energy is of a type. Base load running all day would be expected to get around the average SMP. As can be seen the rate achieved was $58 compared to $62 average. This is caused by low spot price in most of days.

Semi-forced energy generation is required by the daily water flow. This generation has to take place every day but only is selected trading periods. This allows some flexibility in operation enabling the energy to be generated when the prices are higher. The return is usually greater than average rate of return.

There can be some price responsiveness within the day, but this can be limited depending upon the flow requirements.

Figure 6 demonstrates semi-forced generation. It can be seen that the rate achieved was \$83 compared to the average SMP of \$62 and was close to the average peak SMP of \$87. Due to the constraints of operation, some of the generation happened in the lower price times limiting the overall performance.

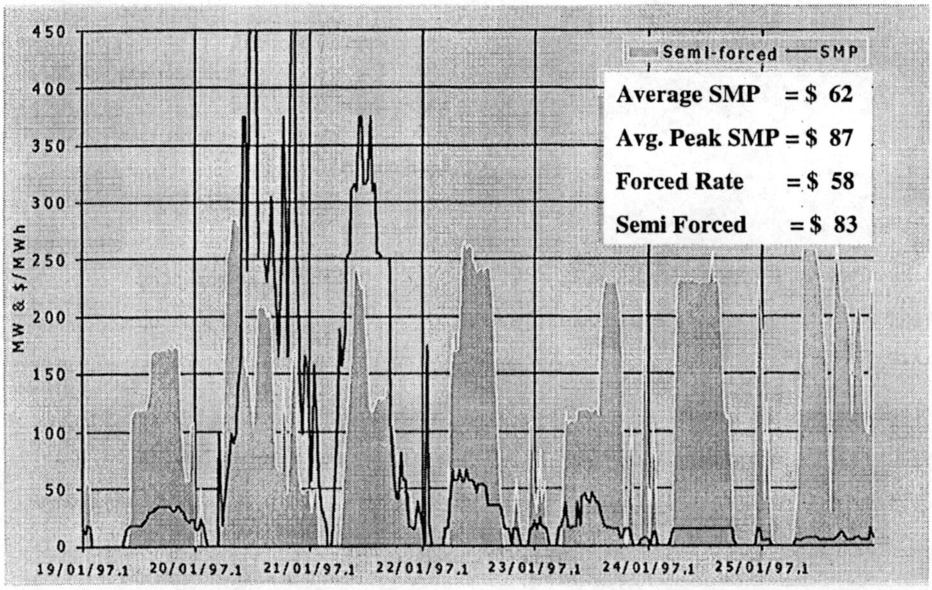


Fig. 6 Semi-Forced generation earning on pool

"Unforced" generation is energy that can be stored for long periods and used when it can provide the maximum return. As seen in Fig. 7, there was a limited amount of unforced generation used during a week. It has returned a rate of \$217 compared to the average of \$62 and the average peak of \$87. The energy from the unforced generation is flexible and can be used to quickly respond to changes in price. The use of this energy can lead to return well in excess of the average SMP prices.

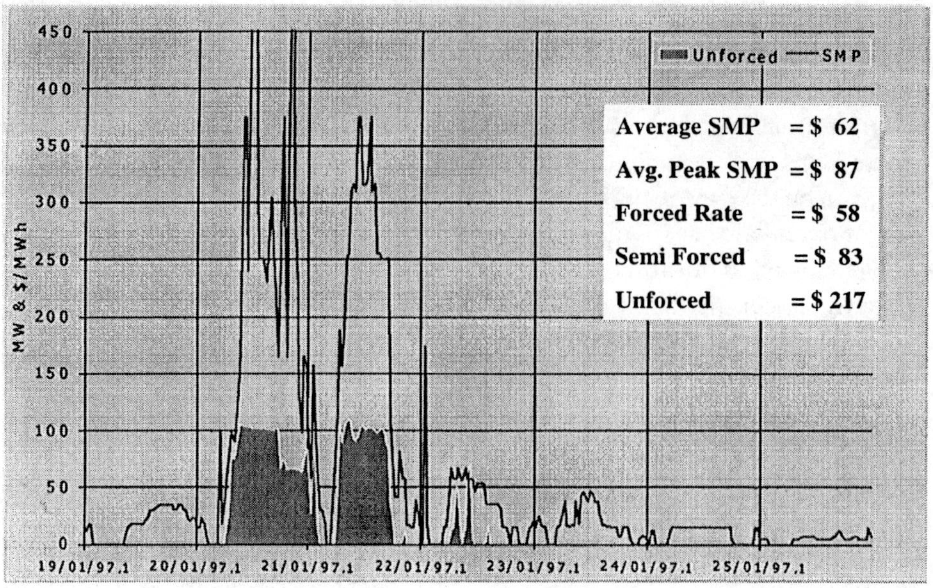


Fig. 7 Unforced generation earning on pool

Figure 8 shows a portfolio of the Southern Hydro generation, which has elements of all three types of generation:

- Base load - Forced ,
- Intermediate - Semi forced
- Peak - Unforced.

The more flexible the plant and water constraints the higher rate can be returned.

| | |
|---|---|
| Average SMP = $62 | Forced Rate = $58 |
| Average Peak SMP = $87 | Semi-forced Rate = $83 |
| Unforced Rate = $217 | |

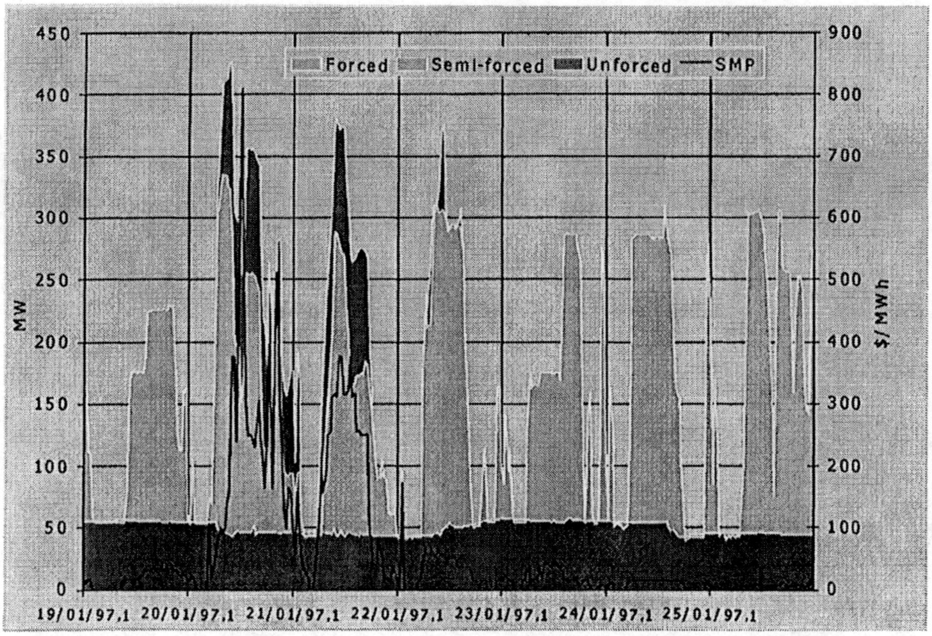


Fig. 8 Portfolio of generation earning on pool

## 5 Market Risk Issues

A several factors have impact on the market risk issues. They include:

1. **Generation:** Electricity generation in hydro power stations dependents upon weather and may vary with the arrangement of water flow schemes. Dry weather may increase the use of irrigation schemes generation but reduce other schemes. Conversely, wet weather may reduce or increase irrigation schemes (depending upon reservoir levels) and increase other schemes. Generation also depends upon reservoir levels. Plant Breakdown will affect all generators.
2. **Pool Price:** Pool prices vary considerably during trading intervals. They depend on demand and excess of generation but also on behaviour of market participants. Low spot prices caused by strong competition mean lower revenue for electricity generating businesses.
3. **Contracts:** When a power generating company contracted significant amount of its production, the company is exposed to the market when the prices are high. If during trading intervals when spot price is very high a generating company does not provide electricity to the pool, it does not receive payments from pool, but the company will have to compensate high spot price due to the contracts. This exposure can be controlled with generation revenue to offset contract revenue, but if generation is not

available, due to water or plant then high losses could occur. These risks can be offset to some degree if there is a very liquid contract market, but the cost to reverse a position may still be costly.

4. **Capacity Effects:** Another problem to consider is the affect of extra capacity. If more plants are offered to the pool the price stays the same or falls. If a smaller number of plants are offered, the spot price usually rises. If generating company wants higher prices one of a possible way is to remove a plant from a bidding offer. This of course may or may not lead to increased revenue, it will depend upon contract position, energy volume contracted, and the contract strike price.

Figure 9 shows the impact of the reduction generating capacity available on spot prices. A lack of generation and transmission line capacity limitations are two main reasons of spot price volatility.

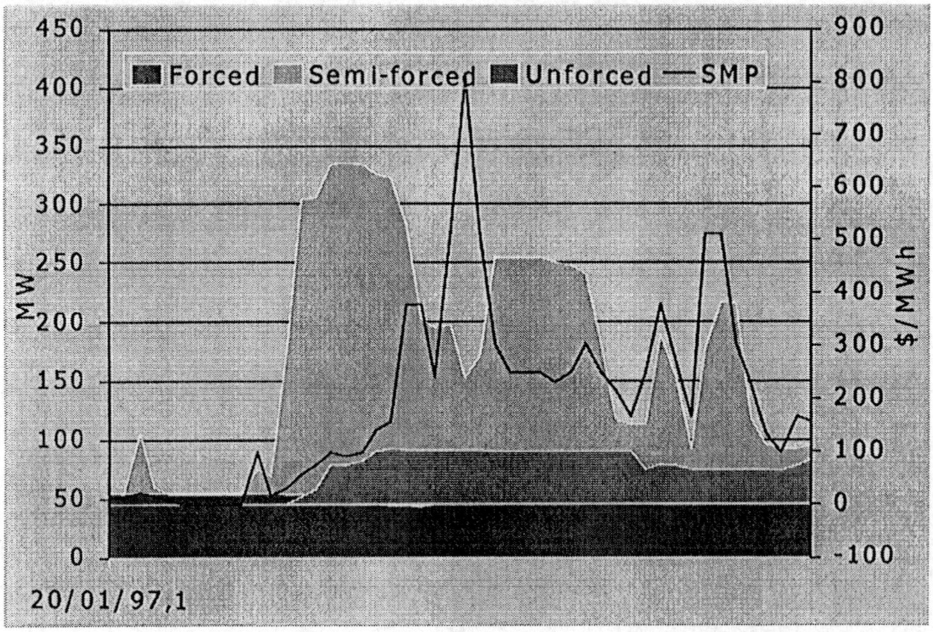


Fig. 9 Capacity effect of extra generation of pool price

The past experience shows that strong competition among generating companies usually drives the prices down. A key analytical tool for any pool or contract trading is forecasting of pool prices. Such prices cannot be predicted precisely; however some scenarios can be constructed to deal with market risks. The scenario underlying assumptions can be tested using Monte Carlo models and other techniques.
Pool prices change quantum levels due to changes in the following inputs in the market:

- Bidding strategies

- Generation availability
- Demand changes.

Volatility usually occurs with low reserve levels or where the pool price is set around the "break points" in bidding strategies. Although VoLL has a low probability of occurrence, but it causes high risks to all market participants. It is likely that the current level of VoLL equal to $5,000/MWh will be changed to $25,000 or$30,000 /MWh.

Another factor affecting market risk is market suspension, and or involuntary load shedding of customers (domestic load). Currently, the National Electricity Code Rules are not clear about how the market will be priced in such cases addressing the issue to the National Electricity Code Administrator (NECA) for a decision.

When a generating company deals with risk management, it has to consider the following issues:

- How much do I expect to generate?
- How much do I expect to earn?
- What is the probability?
- What are the risks?
- How does this match generation & contract position?
- What are the different scenarios and what is the probability that they will occur?

For example, VoLL could occur for the whole month, but the probability would be almost zero.

The consequence if the scenario did occur, is high prices leading to high revenue in the following sectors:

- Pool Revenue
- Contract Revenue
- Option Premiums

If prices are high then pool revenue will be high. Last summer, there were only few trading periods when spot prices were above $300/MWh but in one day, spot prices reached nearly $5000/MWh. If contracts are designed to provide a certain revenue stream, then they cannot take advantage of high prices. In a case of one way contracts, a generating company needs to receive the appropriate amount in premiums for the strike prices to compensate the risks. If there are high prices and a company loses generation then high exposure is faced by the business.

The contract position taken by the hydro generation business can be considered in terms of the type of generation that is being contracted. The following summarise the types of generation contracting are considered:

- "Forced"/Base load generation at expected to receive market prices.
- "Semi-forced"/Intermediate load generation should be offered at peak market price.
- "Unforced"/peak generation more suitable for caps/options and opportunity ie. when spot prices reach the maximum.

There are some hints that can be useful for less experienced market participants:

- Never contract all capacity for all the time, as the type of generation is energy limited.
- Also never be fully exposed to the spot market. With high price risk, only contract at expected pool prices, or better with flexible energy, reduce market risks.
- Risks need to be shared in contracting. If VoLL is unlikely to occur and some retailer wants to pay reasonable option fees for covering the risk of VoLL, then maybe the number of VoLL events covered needs to become limited.

## 6 CONCLUSION

Hydro generation is flexible, and it has the ability to rapidly respond to demand/supply imbalances at any time. However, energy limited generation results in a low (25%) capacity factor. The generation still can earn premium returns, but not all of its energy is as flexible or available all the time.

Hydro generation can take market risks due to its flexibility. This can also be used to capture high premiums the system is above normal operation levels.

Southern Hydro actively participates in market development, in all issues that may affect the company market position. There is a lot of involvement in forums that consider the following issues:

- Electricity code
- Submissions to ACCC
  - ♦ re-bidding - probably limited, but with notional stations then can still move energy in short time frames
  - ♦ need to ensure notional stations in National
  - ♦ spot price - need to keep under review - unlikely to change in short term
  - ♦ VoLL likely to increase to 15-30,000 - increased premiums and returns.
- SFE (new products, options and more liquidity, better risk management)

Southern Hydro has implemented Monitoring & Management System. This system allows:

- Twenty four hour trading – monitoring, changes and response

- Analysis of trends and competitors
- Co-ordination of maintenance and other activities.

Southern Hydro generation has had stable revenue. The contracts have provided funds for expenses before Interest, Tax & Depreciation (EBITD). The performance was very good. During past years the average price obtained was 43% above the average spot price. The performance can vary with water reserve available and volatility of pool prices. The company is looking at using a Linear Programming based package to improve optimisation of generation over the long and short term.

New opportunities being considered by the business include:

- Increased contract presence.
- Development of new marketing instruments.
- Increased trading, depending upon risk profile.

We have developed from green fields to an active market team, positioned to respond to and take advantage of market opportunities, looking forward to challenges and opportunities in the coming years.

****** 9 ******

# Competitive Electricity Markets A Retailer Perspective

**Peter Smith**
Acting General Manager, Energy, Eastern Energy Ltd, Melbourne, Australia

## 1 Introduction

This short paper will focus on the effects of a competitive market for Electricity Retailers. In order for the paper to have some resonance with Retailers elsewhere in the world, the reference to the peculiarities of the Australian Electricity Market such as ownership and regulatory regime, will be kept to a minimum so that more general lessons can be drawn from the market experience to date. It is important to note that the architects of the current market design learnt a great deal from experiences in the UK. Chief amongst these lessons, is real wholesale competition is the driving force behind successful retail competition (if success is judged by size of the discounts off pre deregulation tariffs). However, in addressing these issues, new problems have been raised.

## 2 Background

Electricity supply in most states of Australia up until 1994 was dominated by a large, vertically state owned monopoly. This situation has been rapidly turned on its head with eastern and southern states deregulating firstly their own markets, and finally working together to create a single national electricity market (NEM).

This market and precursor state markets have two key components, a wholesale market for electricity (mostly referred to as the "pool") and an open access regime for the transmission and distribution networks.

The guiding objectives of NEM are:

- freedom of choice for electricity buyers;
- non discriminatory access to the interconnected transmission and distribution networks;
- merit order despatched based on bid price;
- no discriminatory legislative or regulatory barriers to entry for new participants in electricity generation or retail supply;
- no barriers to interstate or intrastate trade and uniform and cost reflective transmission pricing.

## 3 IMPACT ON RETAILERS

### 3.1 RISK MANAGEMENT OR SPECULATION?

The impact on the retail parts of the formally vertically integrated monopolies has been enormous. These changes have affected both the privatised entities as well as those that remain in state ownership. I will focus on the common effects ignoring the tremendous impact privatisation has had. In nearly all cases the deregulation has been preceded by desegregation. This has spilt the former monopoly into transmission, generation and combined retail and local distribution companies. As noted above, the paper will focus on what affect the competitive electricity market has had on retailers noting in all cases these retailers are allied with a network business.

The competitive market has led to retail margins falling to levels not considered possible prior to deregulation. Those customers who have had a choice (the deregulation is phased) are currently enjoying prices that are not attractive to new entrants. These prices barely recover variable costs associated with the end-to-end supply of electricity. Many commentators have noted that this is unsustainable, however most companies engaged in the market share tussle are owned by large overseas utilities and derive most of their income from the monopoly network business hence cutthroat competition could prevail for some time yet.

There has been considerable focus on energy trading as most retailers believe that true competitive advantage can only be achieved by purchasing energy more cheaply on average than ones' competitors. This cost competitiveness has been realised by a number of companies in the Australian market unlike the situation in the UK where retailers were confronted with a duopoly. True wholesale competition has delivered a price that is more random rather than one, that was subject to manipulation. Retailers in our market have not been punished for taking pool exposure but have been rewarded handsomely, although most of the benefit has been competed away with discounts to customers.

This of course has led to some interesting developments in how shareholders view risk in this market. The traditional shareholder has viewed an investment in a utility as safe, offering a low but secure return.

There seems to be a new breed of utility investor willing to take on considerable risk. These risks are not insubstantial given the immaturity and lack of liquidity in the emerging electricity market. These companies believe given true wholesale competition, calculated gambles are worth the risk if significant competitive advantage can be gained. This strategy has worked in a falling market and it is possible that customers will remain with these retailers once the large premiums disappear. These retailers expect the churn rate to be considerably lower when these customers come up for re-tender and are confronted with discounts of only 1 to 2 percent rather than 20-40 percent when the market opened. It begs the question of course; what will their customers do when if the price rises substantially?

## 3.2 Is Electricity Retailing Close to a Natural Monopoly?

This fixation on energy purchasing has been the primary driver but there has also been a secondary consideration - the cost of retailing. It is possible the focus will eventually move away from energy purchasing as companies realise that risking large amounts of share holder funds is not worth it for the low margins obtainable and a "me too" adoption by all retailers (at least those retailers who have marketing executives with market share as their main performance indicator), of taking large pool exposure. This "punting" in many cases has been glossed up as risk management.

However, any objective observer, would note that the recent electricity market has very few of the attributes of financial markets it is seeking to emulate, the most important being liquidity, hence positions while relatively easy to enter into, are relatively hard to exit. Option prices tend to be so high that limiting upside risk is not possible if one is to be competitive with other players who take pool exposure or simpler risk management products.

The focus will again be on reducing the transactions cost of electricity retailing. The largest cost being customer service (call centre, billing, debt management, tele-marketing). Depending on the regulatory regime most of these costs will be required to be recovered from the retail market. Hence retailers are adopting one of three stances. They are looking to outsource, enlarging the retail business or separating the customer service function from the retail business. All three are strategies to achieve economies of scale, scope and promotion.

The savings will be achieved by companies either outsourcing to more efficient customer suppliers or achieving more significant economies of scale by growing very large, possibly 10,000,000 customers or more (if evidence from the United States is to be believed). The ability to leverage or spread the costs over a large number of customers is key, with the important driver being the cost of the customer information system. If there are no diseconomies of scale, it is easy to imagine a single dominant retailer emerging.

This presents problems for the Australian market because there are not 10,000,000 customers in Australia. However, the 10,000,000 customers might be spread all around the world or alternatively customer information systems are used to sell more than just electricity to the customer. The end result will be fewer retailers (maybe one retailer if the regulator does not intercede) and much lower margins in the mass market.

****** 10 ******

# TRADING VARIABLES ASSOCIATED WITH MANAGING A RETAIL LOAD IN THE VICTORIAN WHOLESALE ELECTRICITY MARKET

**STEVE ABBOTT**
Senior Energy Trader, The Australian Gas Light Co. Ltd.

## INTRODUCTION

There are many risks faced by retailers in the Victorian electricity market. Many of these risks are faced by any retail business – customer default risk, legal risks etc. However, because the Victorian wholesale electricity is sold in a traded market, retailers also face trading risks not common to normal retail operations. Many of these risks are similar in any traded commodity market. However, there are also some risks that are unique in some ways to electricity markets and the Victorian electricity market in particular.

This paper identifies three broad types of trading risks identified – load risk, pool risk and contract risk. Within these broad boundaries there are components that can be managed by the retailer and components that the retailer must take on as part of being a retailer. In practical operations no electricity retailer can trade completely trading risk free.

## BACKGROUND

### MARKET STRUCTURE

In November 1994, the Victorian government split the State Electricity Commission of Victoria into a number of corporatised entities. These corporatised entities were broadly categorised into 4 types - a market system operator, a transmission system operator, a number of generation companies and a number of distribution businesses.

The task of the distribution businesses was initially two fold - to operate the distribution network within an allocated geographical region and to retail electricity to customers in that area.

The retailing of electricity requires distribution companies to operate within two markets - the retail and wholesale markets. The retail market exists between customers and those companies with a retail licence[1]. Companies with a retail licence receive income from a customer for sale of electricity and pay "the pool"[2] for purchasing that electricity. This purchase of electricity is effectively one side to the wholesale market.

Initially, the wholesale market existed between those who supply electricity to the pool (generation companies) and those who buy from the pool (those companies with a retail licence and retail customers). From the beginning it was obvious that the pool price was too volatile for normal business practice, with generation revenues and retailers purchases being so dependent on highly volatile pool price. This is illustrated by Fig. 1, which shows the monthly average pool price, since the commencement of the pool.

As a result of this volatility a secondary market for risk management instruments evolved. It is Energy Trading's role to manage these risk instruments, buying and selling forward contracts to best effectively hedged the risk of pool price volatility. This market also attracted a number of non physical market participants whose expertise centred on the management of risk and trading. Gradually the market for forward contracts ("over the counter" or OTC market) has become market unto itself, with participants trading forward contracts for profit as well for the risk management of retail market load. However, the underlying driver for the contract market remains expectations of the pool price.

## POOL PRICE DETERMINATION

The basis of the determination of the pool price lies in the pure market economic theory. In pure market economics inputs of supply are added incrementally from cheapest to most expensive until supply meets demand. When supply and demand are equal, it is the final or marginal unit, which determines the price for the entire market.

In the wholesale market the pool price is determined in this way on a half hourly basis. Blocks of half hourly capacity are offered into the market by the generation units with a corresponding price valued in $/MWh. The generation companies can make up to ten offer/price bands for each of their generation units. The market system operator then forecasts the half hourly demand for the Victorian system and organises each of the capacity offers from lowest price to highest price - the offer stack (colloquially known as the "bid stack"). The highest price in the stack or the marginal unit price, then becomes

1 Initially distribution companies were issues with retail licences but since this time other entities such as generators, other energy companies and other companies that have synergies with retail electricity market have since successfully applied for a licence.

2 The pool was intially a Victorian based pool, hence the name "VicPool". However with the commencement of the national electricity market (NEM) on 24/5/98, this will revert to a national pool.

the pool price. That is the price that is received by all generation units and price paid by all retailers (assuming they have customers). Currently the pool can be set at between $0/MWh and $5,000/MWh, with the cap amount referred to as the Value of Lost Load or VoLL.

The theory associated with this price determination is relatively simple. In practice however, there are a number of variables that mean that the forward determination of this price is very difficult. These issues are discussed later.

**Figure 1 - Average monthly pool price**

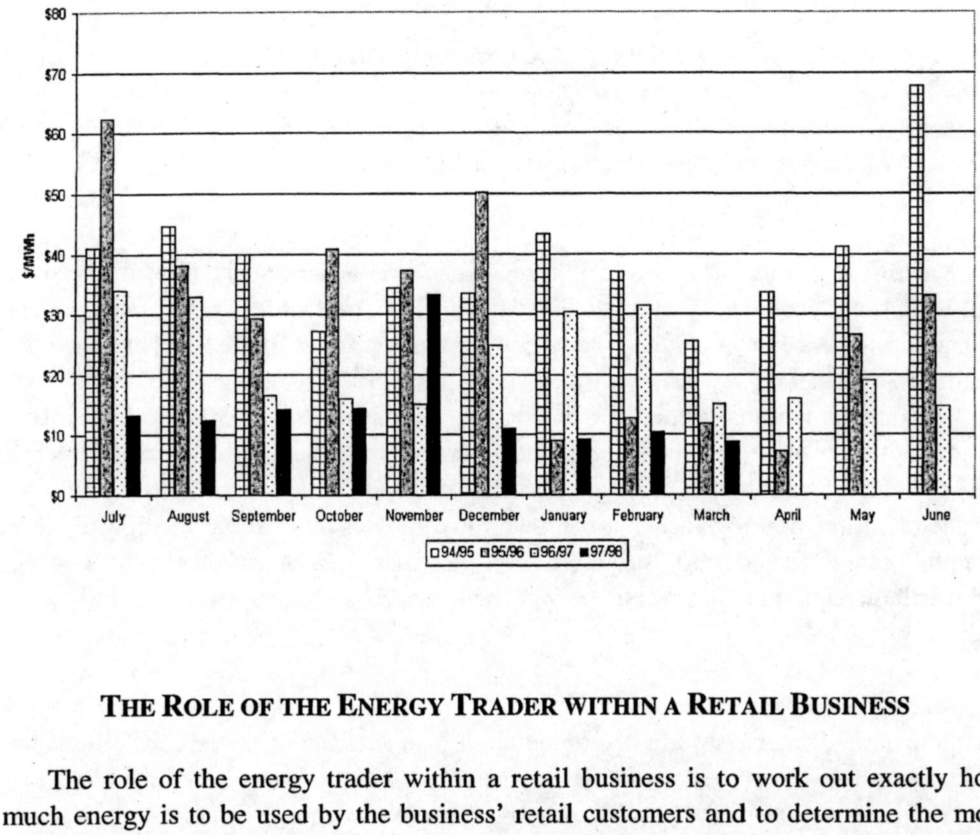


## THE ROLE OF THE ENERGY TRADER WITHIN A RETAIL BUSINESS

The role of the energy trader within a retail business is to work out exactly how much energy is to be used by the business' retail customers and to determine the most efficient way of managing this load in the wholesale market. In a "pure" retailing business that focuses entirely on retailing to end use customers, the role of the energy trader is one of risk minimisation - to perfectly hedge the retail load at a price which is less than the price sold to the customer. The difference between these prices becomes the profit of the retail business.

There are two alternatives to this pure risk minimisation scenario, both of which are more risky but have potentially larger returns. The first is a situation where all or part of the retail load sold to customers is unhedged or subject to pool price. Since the commencement of NEM 1, the contract market has traded at or above the pool price.

This differential reflects the cost of the risk premium. Given that this risk premium exists, the optimal solution for a retailer would appear to be to remain unhedged. However, the uncertainty and volatility of the pool and the risk adverse nature of retailers has meant that this premium has continued to exist.

It is also possible for energy traders to increase their risk and possible returns or losses through taking positions in the forward market prior to the commencement of the contracts - conventionally known as "long" and "short" positions in commodity traded markets. This type of activity has recently been accelerated by the increase of non-physical market participants in the wholesale market.

## TYPES OF TRADING VARIABLES

There are broadly speaking three types of pure trading risks that a retail business is exposed to - load risks, pool price risks and contract market risks.

### LOAD RISK

Load risk refers to the ability of the retail business to accurately measure its own retail load. Retail load can be estimated by summing the loads of retail customers. Loads for these customers can vary due to the size of the business (size of the load) and the way in which the business uses load over time (shape of the load).

For example, you may expect a manufacturing firm may have a large load between the hours of 9am and 5pm on week days as compared to a street light which will use a relatively small amount of load between dusk and dawn. By adding this expected load and shape information together you can determine the customer's expected profile. By summing these expected profiles, the retail business can gain an insight into how much load it will be required to manage in the wholesale market - a total portfolio profile.

#### *Determining retail load*

There are 2 primary ways to determine retail load - by use of historic metering data and by consultation with customers. History has proved that historic metering data is the most accurate measure of how a customer is going to use energy in the future. Recent developments in metering have meant that a retail business can measure a customer's energy usage on a half hourly basis. This not only solves the problem of how much energy a customer is going to use but also the way in which the customer uses energy during the course of the day.

However, the cost of this sort of metering (and related communication and data storage) is quite expensive - although improved technology is constantly reducing these costs. At this stage it is uneconomic for such metering to be installed for smaller customers. These types of customers normally have the traditional metering that measures one total energy figure at the end of a period. This gives the retail business some insight into how much energy the customer uses but no information as to the customer's profile over the period. In this case a standard load shape is used to create a profile. This standard load shape can be determined through sampling procedures on these types of customers. Once the load shape has been determined it acts an assumed shape for that customer. One analogy is that a standard load shape is like a jelly mould with total measured energy being poured in to form a profile of energy usage over time.

The other form of load determination is through discussion with customers. This can be of use either when no metering information is available for these customers (ie..new retail customers) or when the customer envisages changes to the nature of its operations (and therefore changes to its load). Usually some form of billing information is available.

The underlying reason behind why load risk occurs is because it is forward looking - an estimated profile. A total portfolio profile is dynamic in nature in that it can change from expectation for a number of exogenous and endogenous reasons. Endogenous factors include those changes that occur within a given set of customer loads. Exogenous factors include changes in the customer base, which will also impact on the total portfolio profile.

***Endogenous factors***

Endogenous factors occur because a customer's load is different from expectations. These errors can occur for the following reasons:

1. The past metering information does not reflect any underlying growth associated with the customer's operation.
2. The past metering information does not reflect changes to the nature of the customer's operation.
3. The load shape associated to determine the profile for this customer was misallocated.
4. The load shape associated to determine the profile was correctly allocated (ie. best fit of all profiles) but did not fit the profile exactly (all customers are a little bit different).
5. Retail loads are sensitive to weather to different degrees. Different weather conditions can change the nature of profile of retail load significantly.
6. Customers are able unable to determine with accuracy issues of new load - how much the load will actually be, when will the load be fully/partially required.
7. Customers are usually able to predict growth in load in advance but are not able to determine decreases in load associated with closure etc. Even if they are able to determine reductions in load, historically they are not willing to suggest it.
8. Customers perceive economies of scale price benefits associated with load and therefore are more likely to overstate their load than to understate it.

These reasons in combination result in a degree of residual error existing between estimated and actual loads. Historically, estimated profiles have tended to overstate actual loads for reasons 6, 7 and 8, with diversification of errors (both over and understating) resulting in reasons 1 to 5 tending to offset one another.

In order to solve some of these problems monthly pool settlement information has been used to verify profiled information. Monthly settlement information is the most accurate representation of actual retail load. For example, an expected retail profile is created for a calendar year. January's pool settlement shows that this profile is between five and ten percent overstating actual load. This information can be used to correct the profile to better fit what will be expected for the remainder of the year, with allowances made for expected changes in customer load.

***Exogenous factors***

The above discussions of load variability have assumed a stable customer base. However, in reality the customer base of a retail electricity business is constantly changing. In the Victorian experience this variability has increased, as more customers become contestable and as customers become increasingly aware that they choose the length of their retail contract. As well as the necessity to reassess the retail side for the business on a regular basis there are a number of risks associated with changes in the customer base.

Any forward retail load estimate over and above currently contracted customers is purely hypothetical. Based on the assumption that price is the prime determinant in choice of retailer, it will be difficult to know in advance how successful the retail business is going to be in gaining incremental load because it depends on the competitiveness/approach of other retail businesses. Therefore any forward buying in the wholesale market in expectation of retail load has some risk attached.

Most of the movement of contestable customers between retailers occurs on the anniversary of dates of contestability. At this time settlement systems struggle to cope with this rapid movements. An approximate calculation is therefore used as an interim procedure, with more accurate settlement calculated over time. This approximation has a higher degree of load error.

There is a risk associated with where the sale to customer is recognised from the perspective of the wholesale market. Final contract negotiations may evolve over a long period of time. To avoid time risk a sale may be recognised by a letter of intent from the customer. This letter may be the trigger for purchase a hedge for that load in the retail market. However, the letter may not be a legally binding document and if the contract negotiations break down, the retail company may be in a situation where they have hedged without a corresponding retail load.

The nature of a retail/sales based business is that won accounts are well known events whereas lost accounts tend to not be as well publicised. This results in a bias of over stating retail load.

In the Victorian experience, most customers have tended to sign/re-sign on a financial year basis and the period of contracting tends to be just before the beginning of this period (May and June). As a result the wholesale load out past the current financial year is uncertain until just before the beginning of the new financial year.

These risks suggest that, although the degree of uncertainty of customer base tends to focus at the beginning of the financial year, the constant churn of customers from one retailer to another has meant that retail load remains a moving target which requires constant management. Quite a deal of time and effort of retail businesses been put into developing systems and procedures in the retail market to minimise these risks.

However, the greatest risk associated with retail load determination both on the profile and customer side remains human error - both on the customer and retail business side in determining their load. The newness of the market means that individuals do not have the understanding or the historical information to ensure that the information is accurate. It is expected that on one hand that retail load determination will improve over time. On the other hand, domestic contestability set for December 2000, will provide its own new set of challenges on this front.

**POOL RISKS**

Pool risks are the most volatile of all risks faced by a retailer. Actual pool price can vary from expected pool price because of weather driven changes in demand, changes in generator bidding strategy, unexpected generator outages and changes in the regulatory regime that governs the pool. There is also a relatively newly created issue in inter-regional risk.

Retail companies are exposed to pool risk when retail load and hedge contracts are not perfectly matched. Given the uncertainties associated with retail load determination as outlined above, it is near impossible for retail companies to not have some exposure to the pool. The degree to which the pool is greater than the effective energy price sold to customers is the degree to which the retailer is losing money.

As outlined previously, the pool price is determined by the combination of generator supply offers for a given level of system demand. The way in which a generator offers will depend upon its portfolio of hedge contracts. The greater the proportion of total hedged capacity to total capacity, the less effective exposure the generator has to the pool price.

Historically, generators have had most of their load hedged. In this situation, the main issue for the generators is to be dispatched in the offer stack so as to ensure they have equivalent revenue from the pool to match the payment they have to make to the retailer under hedging arrangements. Given this and the over supplied nature of the Victorian electricity market, the pool price has been very low. A very small proportion of generation, the unhedged component, has tended to be bid in at very high prices (at or near VoLL) so as to maximise returns where it is required during periods of high demand. As a result the offer stack (price on the y axis and demand on the x axis) has tended to be exponential in nature, with changes in the stack due to fluctuations in bidding strategies theoretically due to changing patterns of generation hedge portfolios.

Figure 2 - Maximum temperture vs. maximum demand

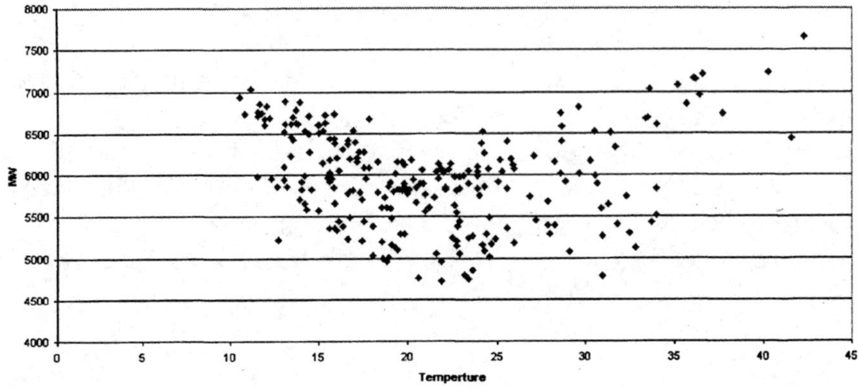


As well as playing a major role in retail load determination, weather also plays a major factor in pool price determination. Days of extreme weather effect a retailer's load, they also effect the total system load. This demonstrated in the V nature of Figure 2 - sensitivity of demand to temperature. When the total system load is greater than expected system load, higher priced generation bids must be called on to meet this new increase in demand - effectively pushing the price up. Historically it has been the days of high demand that have driven the pool price to its highest levels for this very reason.

In theory, unexpected generator outages can also drive up the pool price. Due to the immediacy of these events, a generator may wish to change its bidding strategy immediately to effective withdraw the unit's capacity from the offer stack. The withdrawal of one of Victoria's largest units (500MW) would be equal to a losing approximately 8-10 % of average demand in Victoria. This should effectively mean that higher priced offers would be called upon and the price should rise.

However, the current oversupplied market and resultant flatness of the bid stack, has meant that even if 500MW were to be withdrawn, other generators would effectively take up the slack without a significant change in the price. This is obviously dependent on the speed with which these other generators change their offer strategies. It is expect that the importance of outages in the market will play a greater role in pool price determination once demand grows sufficiently to soak up some of the current excess supply.

Given the new nature of the Victorian electricity market, its regulatory regime has been constantly evolving. To the extent that this evolutionary process affects the pool, it also creates a substantial degree of risk for all participants making forward looking decisions. A prime example of this was the introduction of NEM1 in April 1997, an effective new regulatory regime that joined the NSW and Victorian pools together.

Without discussing the details of such a move, this effectively changed the fundamental determination of the Victorian pool price. Proposed changes to the level of VoLL and the zero pool cap and the inclusion of ancillary services in the price algorithm are examples of future issues that could impact upon the price if and when they are introduced.

A final point on the issues of the changing regulatory regime is that although the goal of pool's regulatory framework is to provide an efficient market that sends out the correct price signals to participants, there is also the issue of system security or "keeping the lights on". The extent to which this is the priority of the electricity market impacts upon the underlying results of the pool. It also brings in a number of external parties (non wholesale pool participants) whose focus is not primarily on what impact regulatory changes may have on pool outcomes.

The final uncertainty associated with the Victorian electricity market is the concept of inter regional risk. Just as in Victoria, a wholesale electricity pool exists in New South Wales with a transmission inter connector between the two states. Since the introduction NEM 1, the offer stack has been determined on a two-state basis to the extent that the cheapest load is called into the stack first, regardless of which state it comes from. In theory, the marginal or price setting bid in both states is the same[3].

However, in reality there are two reasons why the pool prices in the two states could be different. Firstly, there is the issue of the transmission losses associated with getting the electricity across the state border. The value of these losses is factored into bid prices. This means that the state that is exporting electricity will usually have a lower marginal offer (and therefore price) than the state which is importing.

Secondly, there is the issue of transmission constraints. Interregional electricity flows are limited by the capacity of the interregional transmission system. When this system hits its maximum and becomes constrained, the higher priced bids of the importing state will be dispatched. Therefore the constraint acts as a barrier to a pure market outcome. When this occurs price differentials can become quite significant. During the days of November 25 and 26 1997 the inter connector became constrained flowing into Victoria, resulting in prices as high as $4814/MWh. In that same time period New South Wales recorded $145/MWh.

The risk that this inter regional situation creates can be quite significant. A retailer takes on this risk when it retails to a customer who is in one state and effectively backs this customer with a hedge in the other state. The retailer makes pool payments based on the pool price in one state but receives pool payments from its hedge counterparty based on the pool price in the other state. Therefore, when the price in the first state is greater than the latter the retailer is effectively losing money. The contract market should effectively reflect this differential. However, because the nature of the constraint issues is difficult to determine, the risk is to some degree random because it is dependent upon the supply and demand issues in both states, which in turn are dependent upon the risks outlined above.

**CONTRACT RISK**

The price of forward contracts is effectively determined by pool price expectations. Therefore as pool price expectations change so do contract prices. This change in contract prices effectively constitutes the contract risk.

Working on the assumption that retailers are looking to sell to their customers at a rate that covers their purchase cost, retailers will price to their customers based upon the rate at which they can buy hedge contracts. However, it may be a significant time between when a customer receives a price from its retailer and when it chooses to contract with that retailer. This creates contract risk for the retailer.

Because of the above phenomenon, it is likely that at some stage a retailer will be "short" at some stage (ie. have a retail load that is not specifically covered by a hedge contract). The risks associated with a short position are the same in any market - if the price goes up the retailer loses. There are two alternatives available to the retailer in this situation. The retailer can contract at the higher price, effectively taking less profit. Conversely, the retailer can choose not to contract. Not contracting will effectively mean that the retailer's purchase price will be subject to the pool price. Contract prices go up because participant's expectations is that the pool will be higher, this could potentially be a dangerous ploy.

Another type of contract risk applies when a retailer decides to buy forward. This means that the retailer buys a hedge contract in the expectation of winning retail customers. If the hedge contract price goes down the retailer is effectively left with a contract that it cannot sell for profit[4]. In this situation the retailer has three options. First, the retailer can discount its price to the customer to effectively lock away a loss. Secondly, the retailer can remain over covered (ie... contract cover greater than retail load). As the contract market has historically traded at a premium to the average pool price over corresponding periods, it is likely that retailer will also lose doing this. Finally, the retailer could look to sell the over cover as a contract in the traded market. As the trading market works on a spread basis, where the sell price is less than the buy price, this is likely to mean that the retailer will have to effectively lock away a loss. In each of the three cases outlined above the retailer is locking away a loss – it is a case of choosing the alternative that minimises this loss.

## SUMMARY

There are many pitfalls associated with retailing in the Victorian electricity industry. The extent to which the retailer has exposure to risks will be dependent upon the quality of business systems to manage load risk and the ability of individuals to analyse pool and

---

[4] Working on the assumption that the lowest price in the retail market wins the customer. This has historically been the case.

contract risk correctly. It will also be dependent upon the amount of risk the retailer is prepared accept. As well as this there will be a certain amount of risk that will be out of the control of the retailer, such as some changes in the regulatory environment.

Trading risks are more significant in that electricity retailing is not proving to be a high margin business with low switching costs and significant competition for customers. Increasing competition has reputedly forced some retailer to reconsider their approach to risk – taking on more trading risk in the wholesale market to survive in the retail market. It is expected that eventually some balance of acceptable risk and expected return will be established to ensure electricity retailing will continue to be a viable business in the future.

## ****** 11 ******

# 'VISION' ELECTRICITY MARKET SIMULATION MODEL

**G.W. KIPP** **G.G.GAMAGE**
Manager Senior Analyst
Power Evaluation Department, Victorian Power Exchange, Melbourne, Australia

## INTRODUCTION

System Planning and Development Division within Victorian Power Exchange (VPX) [1], has developed VISION [2], a multi region electricity market simulation model for use in studying the National Electricity Market (NEM) which is to operate in the South-Eastern Australian states. This market will be operated by the National Electricity Market Company (NEMMCO) [3] under the National Electricity Code [4], which will be administered by the National Electricity Code Authority (NECA) [5].

The line diagram in Figure 1 below shows the South-Eastern Australian regional generation network. Existing links are shown as solid lines. Proposed links are shown as dotted lines. The link capacities vary widely, with the largest capability lines being those between Victoria and New South Wales via the Snowy region, which encompasses the Hydro generating scheme operated by the Snowy Mountains Hydro-electric Authority.

This paper briefly describes the VISION model in five principal categories:

- Design requirements underlying the model,
- Its principal features,
- Input requirements,
- Main algorithmic features, and
- Output.

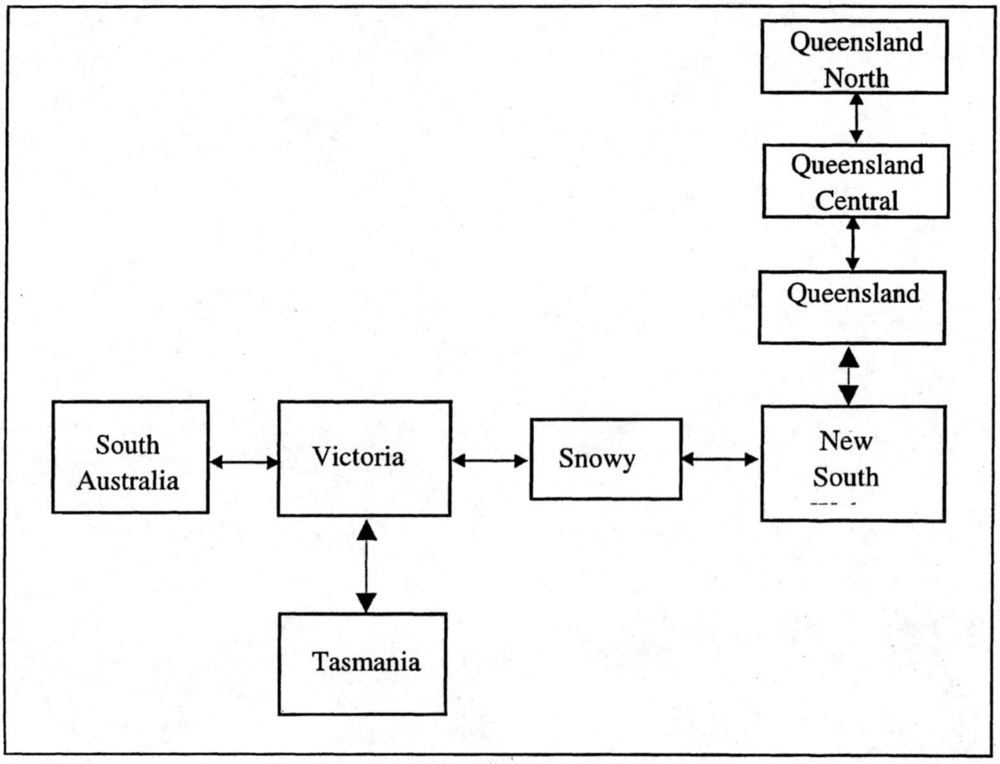


Figure 1. Regional Transmission Structure

## DESIGN REQUIREMENTS

The model was originally designed for internal use by VPX. Because of industry interest by pool participants, network operators, market consultants and other parties, it was made available for public release in May 1997.

Because of the need to respond quickly to market changes, the model has been designed to be:

- Flexible in operation, with the ability to address both short- and long-term scenarios quickly and reliably.
- Representative, with sufficient detail in the modelling of critical market parameters, such as generation scheduling and pool settlement rules, with an acceptable execution speed and readily available hardware.
- Modular, so that market rules can be easily changed to examine different market structures.

- Easy to use, recognising that there will be considerable input and output detail required.
- Stochastic, recognising that generator capabilities and system loads are statistical quantities. This enables the variability of output parameters to be calculated via Monte Carlo simulation techniques.
- Interactive, providing the user with the optional ability to adapt daily bids during the bidding cycle, depending on the progress of the simulation.

  Starting from a scenario-based bid and contract input structure with standardised output reports, the model provides a sophisticated level of capability and complexity within an acceptable response time frame.

## PRINCIPAL FEATURES

The principal features of the model are described below. This discussion covers the operating platform on which the model runs, the key inputs, factors considered in the dispatch and pool price calculations, and output reports.

## OPERATING SYSTEMS

The model is available under dual operating systems. Platforms currently supported include PC/DOS/Windows and Digital AXP/OpenVMS. A common body of code is used for the calculation engine compiled from FORTRAN source code, and kernel libraries are used to separate system dependant features. On the PC, a Visual Basic front end is used to improve the user interface.

## SIMULATION CHARACTERISTICS

Monte Carlo techniques are used in simulating the hourly dispatch of generation to meet the system demand, based on a number of deterministic and randomly varying inputs. The principal random components are highlighted in Table 1.

Table 1. Principal Stochastic Components in VISION

| Quantity | Characteristic |
|---|---|
| Generation forced outage and in-service time distributions | Weibull Distribution |
| Demand trace scaling factors | Rectangular, centred on 1.0 |
| scaling of selected Bid prices | Rectangular, centred on 1.0 |

## MODELLED SYSTEM LIMITS

The model provides for an eight region NEM representation, as shown in Figure 1. Up to 200 generation units and up to 35 distribution businesses may be modelled. Up to five continuous calendar years may be simulated on the PC platform; versions for either eight or 25 calendar years capability exist on the Alpha platform.

## REGIONAL DEMANDS AND CUSTOMER LOADS

Regional demands are supplied as hourly demand traces that may be weighted to achieve a composite load trace. Customer loads may be derived from system loads based on historic load traces and fixed weekly customer load profiles, or may be explicitly-specified hourly loads for each customer. Variability of weekly demands may be imposed by random scaling about the forecast value. A model utility is used to prepare demand traces, based on half-hourly actuals or existing demand traces, to achieve a specified annual energy and maximum demand.

## GENERATION REPRESENTATION

Generation is defined by a number of repairable units grouped by station. Each unit has a defined generation rating, owning region and ramp-up rate. Simplified unit commitment is defined by specifying a unit must-run-to-minimum flag.

No distinction is required between generation types (thermal, hydro, etc.). Regional hydro generation may be bid into the market or specified as an optional annual regional MW profile to account for run-of river releases. In addition to outage and in-service distributions mentioned above, the user may specify plant start-up and shutdown dates and fixed maintenance periods for repairable generators.

## CONTRACTS BETWEEN GENERATORS AND CUSTOMERS

Three types of contract are available for selection, viz. One-way hedge, Two-way hedge and Swap (capacity). Each type may be sculptured by day type (Weekday/Weekend) and hour of day. Separate contracts may be defined between generator groups and customers, or the whole market may be fully contracted to analyse vesting contracts.

## GENERATION DISPATCH

The dispatch schedule is based on NEM rules, with a modified merit order dispatch taking account of the effect of marginal losses on marginal prices. Unit commitment may be fixed or dependent on forecast generation reserve. Merit-order dispatch results in faster calculation than mathematical programming formulations, but with the restriction that only a single flow path is allowed between any two regions. LP formulations to overcome this restriction are being considered.

## MARGINAL LOSSES

In the modified nodal pricing model used in the NEM, generation and demand centres are referred to the regional node by static marginal loss factors. Inter-regional marginal loss factors are dynamic values, calculated each hour using regression equations. These equations are determined using load-flow studies; they are a function of regional demands and inter-regional tie line flows.

## MARKET RULES

Various phases of NEM development may be represented. System Marginal Price (SMP) calculations may be based on isolated regions, a transitional market involving Victoria, NSW and Snowy, or the full national electricity market with inter-regional tie-line constraints.

## TRANSMISSION LIMITS

Transmission limits are modelled as fixed values at nominated dates in the simulation. A limited amount of system dependent limit calculation is available for some tie-lines. Separate values are specified for each direction of the tie-lines connecting the regions shown in Figure 1.

## BIDDING REPRESENTATION

Generation offers and demand side bids are made at the unit level. Bids may be presented in either single block format or multiple-block format with up to four blocks per unit defined by MW capability and price. Reference bids may be varied with a minimum daily frequency. These reference bids may then be adjusted by one of the bidding strategies discussed below.

This representation is a simplification of the NEM bid structure, which allows up to ten blocks per unit and allows changes to block MW availability during the bidding day.

The model representation is regarded as a satisfactory compromise between speed and detail.

## BIDDING STRATEGIES

For nominated generation units, the user may choose to vary a reference daily bid price using one of a number of strategies. The strategies currently available vary the bid price based on random fluctuations, time-of-year, forecast generation reserve and revenue optimisation. An hourly scaling of reference daily bid prices is available as an input option.

## GENERATION TARGETING

Annual generation targets may be defined for any station. The dispatch algorithm endeavours to meet this target by moving MW capability between the nominated bid blocks for the station.

## OUTPUT OPTIONS

Numerous formatted reports are available for printing, in addition to text files that may be used for user post-processing. Printed reports are generated optionally by day, month, year and overall simulation, with a choice of Calendar or Fiscal year basis. Generator results (both energy and payments) may be aggregated into contract groups, and group results into portfolios.

## GENERATION OUTPUT

Output tables are provided for generation dispatch, regional imports, exports, total generation, inter-regional tie-line flows and constraints, regional demands and period maximum MW demands. Detailed output is provided when generation shortfalls occur.

## FINANCIAL OUTPUT

Pool payments by customers and to generators are reported, along with difference payments or contract adjustments. One-way, two-way or swap contracts may be defined between customers and generators or generation groups, with contracts sculptured by time and day type. If fuel costs for generators are available, a report also calculates annual generation production costs.

Inter-regional (IR) surpluses are broken down into marginal loss surplus, constraint loss surplus and average loss deficit components. An input option also allows output based on optional contracts with an IR trader.

## Reliability Output Options

For reliability analyses, a number of output tables are available. These include:

- Loss of Load Probability (LoLP), expressed as the proportion of simulations with generation deficiency,
- Unserved Energy at several shortfall levels,
- Hours of Load Shedding at several shortfall levels,
- Energy at Risk at several reserve levels,
- Load Shedding Severity frequency distribution, and
- Daily Forecast Reserve frequency distribution

## SMP and Generation Analyses

Statistical analysis of SMP results is provided. Distributions of frequency of stations setting SMP and regional SMP and SMP duration tables are provided. Average hourly and daily SMP is also analysed by day type and for user-defined periods (such as peak/off-peak).

Additional output data files are generated for further processing if required, using products such as SAS or Excel. These data include time histories of generation dispatch and line flows, and regional hourly SMP values.

## Operation Modes

Normal operation displays minimal terminal output. A batch mode allows the terminal display to be captured in a log file. A gaming mode option allows the user to vary bids interactively at regular intervals, dependent on simulation results.

## Supplementary Output

A range of diagnostic output allows program operation to be checked at several levels of detail. Comprehensive error, warning and information messages are documented with suggestions for user investigation and resolution.

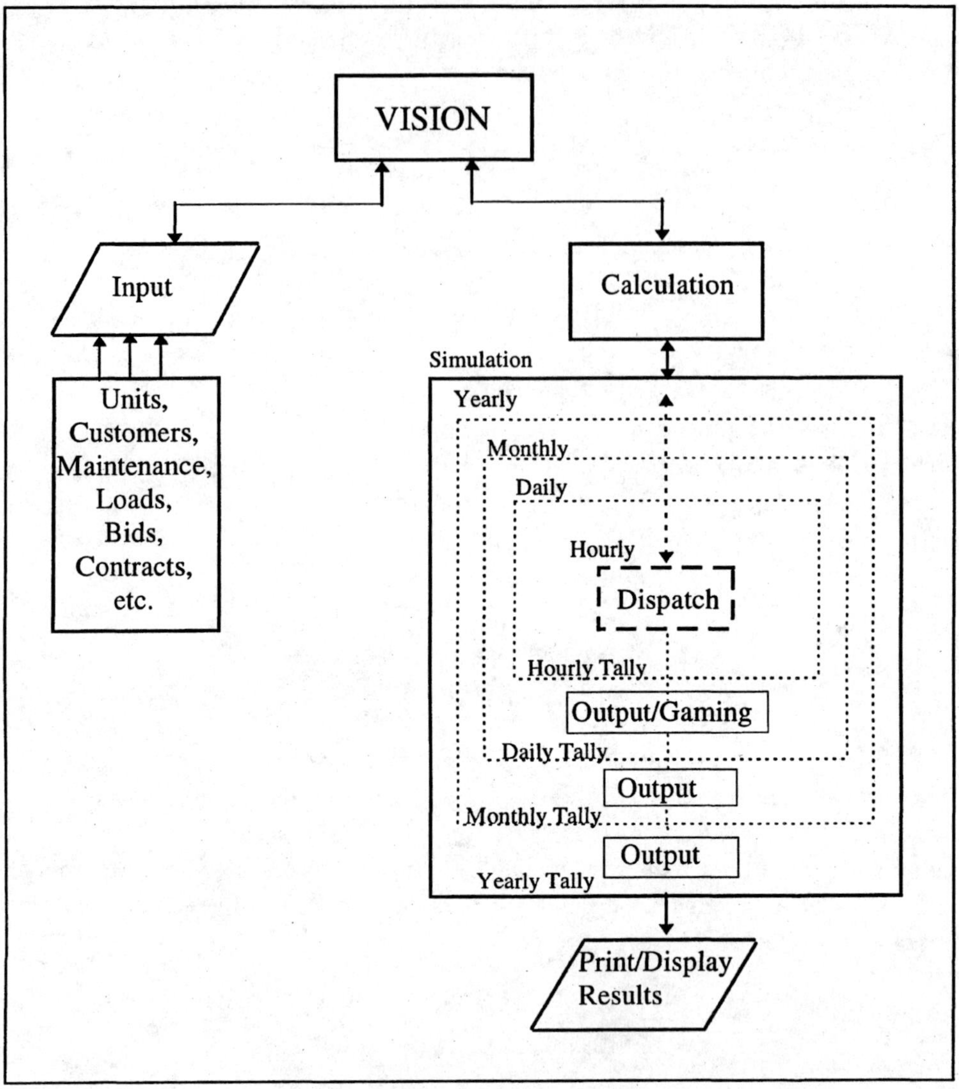


Figure 2 VISION Model Structure

## CALCULATION ALGORITHM

### PROGRAM STRUCTURE

The essential program structure is laid out in Figure 2 below. The three basic components of input, dispatch calculation and output are highlighted and described in subsequent sections.

### INPUT DATA

The branch identified as *Input* in the Figure 2 highlights the fact that all data input is completed before the simulation commences. Input data definition is consolidated in a control file, representing one simulation scenario. A sample input control file with key input data is shown in Figure 3.

Each line consists of a data directive followed by one or more values. The control file may be generated by a text editor or created by the VISION Windows Interface. The principal data inputs provided by these directives have been discussed in section 3 above.

### DISPATCH CALCULATION

The branch identified as *Calculate* groups the calculation components and links the input and output phases. The system simulation for the nominated period is repeated as often as specified in the control file. Within this simulation loop, there are daily and hourly nested loops.

Each hour, the basic dispatch parameters (bid prices, unit status, etc.) are initialised, and a merit-order dispatch is calculated. The detail of the dispatch, and the level of adjustment to the basic dispatch to generate economic transfers, depends on the user input options.

```
BID_STN   =LOCAL:tstn.pro            (generation bid profile)
CONTRACTS =LOCAL:contract.dat         (Contract definition file)
CUSTOMERS =LOCAL:customer.asc         (Customer definition file)
START_DAY =19930701                  (Simulation start date)
END_DAY   =19940630                  (Simulation finish date)
FORATE    =LOCAL:forate.dat  (Generator forced outage rate data)
HYDRO.V   =LOCAL:hydro_v.dat     (Vic Hydro generation profile)
JDEBUG    =  0 19930701.11 19930701.11 (Debug level, dates/hours)
JOB.TITLE1=Title1                    (Line 1 of title)
JOB.TITLE2=Title2                    (Line 2 of title)
LOADS.NSW =LOADS:LN_may92.vpl              (NSW load trace file)
LOADS.SA  =LOADS:LS_9205a.vpl            (SA load trace file)
LOADS.VIC =LOADS:LV_9205a.vpl            (Vic load trace file)
MAINT     =LOCAL:maint.dat     (Generator maintenance data file)
MODE      =N                     (National grid mode)
PERIOD.1  =LOADS:periods1.dat  (Generator energy period 1 data)
PLOTFLAG  =F           (True/False to control gen trace output)
RAMP.UP   =N          (Yes/No for generator ramping on startup)
REPORTING =YF       (Daily/Monthly/Yearly/Calendar|Fiscal) freq)
SEED      =123456789               (Random simulation seed)
TARGET.GEN=LOCAL:target.dat     (Generation target control file)
TRANSLIMIT=LOCAL:trans.dat      (Transmission limits data file)
UNITS     =LOCAL:units.dat        (Generation unit data file)
WEEK_FACT =123456789   0.98  1.02(Weekly random load factor data)
```

Figure 3: Sample Control File

### OUTPUT

At the end of each day, output request flags are checked (daily, monthly, yearly, etc.) and output generated as required. This is indicated in the *Output* boxes in the previous diagram. At the end of each simulation, results for the period are tallied. At the end of all simulations, means and variances of output quantities are calculated and summary statistical reports are generated. The volume of printed reports is under the user's control.

If gaming mode is used, at 24:00 hours on nominated days of the week the simulation pauses for the user to view simulation results and modify bids as desired. This gaming cycle continues for the nominated simulation time. Note that this calculation is deterministic, in the sense that only one simulation for the time period is allowed.

## CALCULATION ALGORITHMS

The principal features of the model have been presented in section 3, but it is worth describing some of the calculation algorithms in more detail.

### GENERATION DISPATCH

Each hour, a unit dispatch is calculated. As indicated in Figure 2, this component is the key element of the calculation procedure. The dispatch algorithm is supplied with system demand, unit generation limits, unit commitment flags and unit block bids; start-up costs are assumed to be reflected in the bid price.

Each hour, generation blocks are loaded in order of increasing bid price, taking account of static loss factors referred to regional nodes. Blocks are loaded to meet the combined system demand, taking tie-line limits into account. A first pass loads must-run plant to minimums; a second pass then dispatches optional running plant until the system demand is met or a system shortfall results. In these passes the effects of inter-regional marginal losses are not considered.

After an initial dispatch has been determined, inter-regional loss factors are then iteratively applied to the marginal generation prices in each region. Plant is off-loaded or uploaded as required, until the true marginal plant in each region is determined. If there are no inter-regional constraints, one generator will be the marginal plant for all regions. If any region is constrained by tie-line limits, its SMP will be determined by generation in its own region.

The generation resulting from the dispatch calculation is available as an hourly, daily, monthly and/or yearly summary. This generation balance is then used as the basis for calculating payments from customers to generators using the resulting regional SMP values.

Bid prices of units nominated for dynamic bidding are adjusted at nominated frequencies (daily, weekly, etc.), as are MW capabilities of units with nominated annual generation targets.

## GENERATION TARGETING

An input option allows the user to specify annual generation targets, based on a monthly profile, to be achieved by nominated generation stations. Units in that station are bid in the usual way, and the MW bids of nominated units are periodically adjusted based on the cumulative generation relative to targets. Account is taken of load factors in the period, moving minimal generation from higher cost blocks in an endeavour to maximise revenue.

## POOL (SMP) PAYMENTS

System Marginal Price (SMP) pool payments are based on the hourly regional SMP and generation or load values resulting from the dispatch. These payments are then summed for each hour in the simulation period.

## CONTRACT ADJUSTMENTS (DIFFERENCE PAYMENTS)

Three types of contract have been implemented in the model. The 'spot' or SMP payment has been itemised above. The essential details of contract adjustments (in terms of customer payment) to this spot payment for each type of contract are given below.

*One-Way Hedge*:

Adj = Option fee - contract MW * (spot - strike)

*Two-Way Hedge:*

Adj = Option fee

+ contract MW * (strike - spot) {when strike > spot}

- contract MW * (spot - strike) {when strike < spot}

*Swap Contract*:

Adj = Capability payment + Generated MW * (strike - spot)

With appropriate specification of MW levels and strike price of one-way hedge input data, the difference payment output may be regarded as a premium level that may be used to determine the timing of the entry of new generation into the market.

## TOTAL PAYMENTS

The total payments to a generator or by a customer are found by summing the SMP payments, reliability payments and contract adjustments for that generator or customer.

## INTER-REGIONAL HEDGES

Inter-regional (IR) surpluses that may be used to underwrite IR hedges are calculated, broken down into marginal loss surplus, constrained loss surplus and average loss deficit components. If IR contracts are specified, standard reports are provided for IR contract payments.

## PROGRAM OUTPUT

The basic output is a periodic record of unit dispatch and financial transfers (daily, monthly or yearly). These periodic data are accumulated for each simulation for output summarising specified simulation periods. The option of daily, monthly and/or yearly output may be specified by the user; overall simulation output summaries are always generated.

Requested output for the last simulation is always available. In multiple-simulation runs, means and standard deviations of output quantities covering all simulations are also calculated. Besides these generation and financial summary reports, several other output analyses are available. The several types of report are described below.

### GENERATION REPORTS

For each requested reporting period, the following data quantities are given, tabulated by generator group (generator unit, and group/portfolio if the latter are defined):

- Period (hour, day, month, year).
- Generation, imports and exports by generation group and region.
- Generation capability by generation group and region.
- Inter-regional energy flows.
- Frequency of inter-regional transmission constraints.
- System generation shortfalls.
- Regional loads and maximum demands.

### CUSTOMER DEMANDS

For each requested reporting period, customer demands

### PAYMENT REPORTS

For each requested reporting period, the following quantities are given, tabulated by generator group or customer:

- System marginal price and SMP payments.
- Reliability payments.
- Contract adjustment payments.
- Net payments.

More details of the payment parameters are given on Daily reports, as these reports are readily used as debugging aids. These quantities include hourly SMP, hydro capability, reliability payment rates, etc.

Inter-regional payment reports are also provided, detailing:

- Marginal loss surplus
- Constrained loss surplus,
- Average loss deficit,
- Total loss surplus, and
- IR contract payments (if requested by input selection).

## STATISTICAL ANALYSES

In addition to reporting the means and variances of standard output quantities given above, the program can provide additional statistical analysis of simulation results.

*Time Period Analysis of Generation.* The user can nominate several periods for which the average energy and generation of each generator can be analysed. This is useful in studying the impact of input data that is sculptured by time period.

*SMP Histograms.* Histograms for each participating region are produced showing the number of occasions when:

1. The SMP falls into nominated price bands, and
2. The SMP was set by particular generating units.

*Hourly SMP Averages.* Tables are generated which show the average hourly SMP rates for each day of the week, for each region. The average is calculated both as a time-weighted value (the usual definition for time-varying data) and as a load-weighted value.

*Time Period Analysis of SMP.* For the user-defined time periods, time-weighted and load-weighted average SMP can also be analysed.

Input options allow continuous traces of hourly generation, loads, tie-line transfers and SMP values to be dumped to files for further processing.

## APPLICATIONS

The VISION market model was originally developed as an aid to understanding the NEM market bidding and contract processes. As the market has developed, the model has also proven valuable in a number of other areas. These include:

- Preparation of pool price paths required by background information memoranda during the deregulation and sale of electricity generation and distribution in Victoria,
- Generation of financial positions for generators and distributors under various market scenarios,
- Analysis of inter-regional trading to evaluate the role of an inter-regional trader in NEM,

- Analysis of inter-connector economics required for inter-connector planning studies,
- Energy trading by market participants to develop bidding strategies, evaluate contracts and analyse risk positions, and
- Analysis of regional reliabilities and reliability standard proposals.

Parties interested in the use of the model in these or other areas should contact the General Manager System Planning and Development at VPX[1].

## CONCLUSION

The VISION model, its background and its uses, have been described. The model established itself as an industry standard during the setting up of the Victorian Pool Market. It continues to be used by the industry in the setting up of the National Electricity Market, and continues to be supported and developed by Victorian Power Exchange as the market becomes established.

## REFERENCES AND ENDNOTES

1. Victorian Power Exchange, Level 14, 452 Flinders Street, Melbourne, Vic 3000, Australia.
2. *VISION Electricity Market Model User Manual*, Victorian Power Exchange, 1998
3. National Electricity Market Company, Level 16, 461 Bourke Street, Melbourne, Vic 3000, Australia.
4. *National Electricity Code*: National Grid Management Council, October 1996.
5. National Electricity Code Authority, Level 7, 14 Martin Place, Sydney, NSW 2000, Australia.

## BIOGRAPHICAL NOTE

Graham Kipp is currently Acting Manager Power System Evaluation Department, Victorian Power Exchange. His experience in the electricity industry is in the area of information systems and power system and market models. The VISION market model development was a prime area of responsibility.

**Gnagnabandu Gamage** is a senior analyst in Power System Evaluation Department. His background is in electrical network analysis and power system planning.

****** 12 ******

# CONTRACTS FOR ELECTRICAL ENERGY

**WLADYSLAW MIELCZARSKI AND GRAZYNA MICHALIK**
Monash University, Department of Electrical and Computer Systems Engineering
Melbourne, Australia, Email: Wlad@basil.eng.monash.edu.au

**Summary:** This paper presents the introduction of the competitive electricity market in Australia. The electricity supply industry in Victoria, the first State which introduced open market, is discussed in details. The electricity is trading on the spot market with financial protecting instruments known as vesting and bilateral hedge contracts. The analysis of various contracts and impact on participants is presented.

## 1 INTRODUCTION

There are three levels of electricity wholesale trade in an open electricity market in Victoria:

- Spot trading
- Vesting contracts
- Bilateral contracts (long and short term)

***Vesting contracts*** are financial instruments arranged between power stations and energy retailers by the Victorian Government before the introduction of the open electricity market. They aim at the market stabilization allowing power stations to recover a long-term marginal cost. Vesting contracts, which cover about 50% of pool demand, will last to year 2000.
***Bilateral Hedge contracts***, long and short term, are voluntary contracts arranged between energy retailers and power stations. There are several types of hedge contracts.

Energy, which is not contracted in the above ways, is traded on the spot price market. Spot prices result from the balance between supply and demand in a bidding system. Currently, participants provide their bids in 10 price bands in 0.5-hour trading intervals. The market manager calculates a spot price in 5-minute periods. Vesting and hedge contract payments are calculated in 0.5-hour intervals by averaging 5-minute spot prices.

The structure of energy trading in Victoria is shown in Fig. 1. The precise value of energy demand covered by various hedge contracts is unknown. It is estimated to count for about 40% of the total energy demand. Although the spot price market covers a relatively small segment of the pool, spot prices affect strongly strike prices in the hedge contracts and behaviour of pool participants.

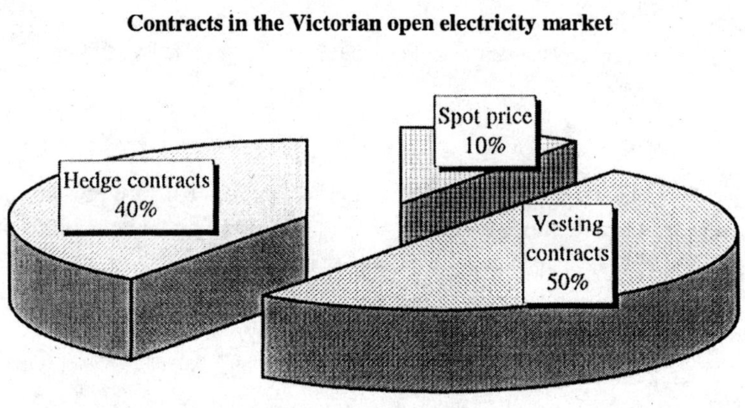


Fig. 1 Estimated share of various types of contracts

## 2 SPOT TRADING

Spot trading operates in the market to ensure that supply and demand are balanced at all times. Generators and Consumers buy and sell energy at the spot price for amounts not covered by the contracts. Generators provide the information on the offered commitment of their units determining trading intervals and prices. The market manager receives the bids and creates a dispatch order in which the plant with the lowest price is run first. Customers can also bid to turn off switchable blokes of loads, called Demand Management units, at a given price. The spot price is set every haft hour by the last Generator selected to run. This price can rise and fall due to the balance between demand and supply. An example of the process of determining spot prices is shown in Fig. 2 and Table 1 [1], [2].

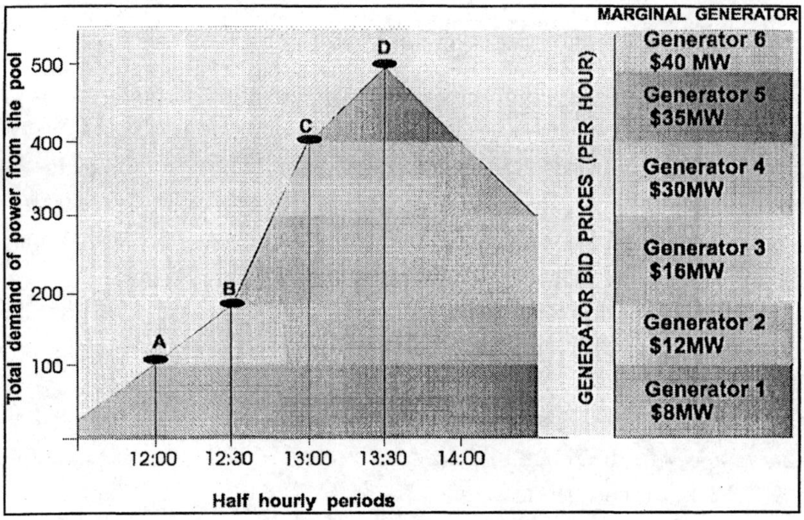


Fig. 2 Determination of spot prices in an open electricity market

Table 1 Spot price determination

| Pool Price | Time | Total Demand | Scenario | Graph Point |
|---|---|---|---|---|
| $35/MW per hour | 13:30 | 470 MW | Generators 1, 2, 3, & 4 are fully utilized. Generator 5 is partly utilized | Point D |
| $30/MW per hour | 13:00 | 400 MW | Generators 1, 2, 3 & 4 are fully utilized. No other generator used. | Point C |
| $12/MW per hour | 12:30 | 200 MW | Generators 1 & 2are fully utilized. No other generator used. | Point B |
| $8MW per hour | 12:00 | 100 MW | Generators 1 are fully utilized. No other generator used. | Point A |

Note that Generator 6 at $40/MW per hour supplies no electricity and receives no pool income for production at this time.

## 3 Vesting Contracts

Vesting contracts were introduced in the beginning of a competitive market in order to reduce price volatility associated with an unstable market. These contracts are financial hedging instruments that provide a field price with reference to variable Pool prices. Vesting contracts have two major components:

- contracting amount of electrical energy for a fixed price to cover demand of non-franchized customers until the year 2000. This is tied up with ***Two Way Firm Contracts*** to cover the differences between the actual spot price and the fixed price determined by the Office of the Regulator General. Two Way Firm contracts are valid only to a strike price which was initially set up at the level of $300/MWh. This value has been currently adjusted to $329/MWh.
- **One Way Non-Firm contracts** that are applied only when a spot price is above a $329 cap to Generator who is selling electricity into the Pool at a relevant time.

One Way Non-firm Contracts are intended to hedge risk of very high Pool prices. These contracts apply to the range of prices between A$329/MWh and the Value of the Lost Load (VOLL), which currently amounts for A$5000/MWh. When a Generator is producing during periods of very high Pool prices, the Generator receives payments for electricity due to spot prices and returns the difference between A$329/MWh and the Pool price.

The ratio of average prices and energy demand contracted in vesting contracts to average values of spot prices and pool demand for May 1997 is shown in Fig. 3. As can be seen, the actual spot prices oscillate around 50% of the vesting contract prices. This is mainly due to the excess of generation and high vesting contract prices. Since power stations are able to recover their fixed costs of electricity production in vesting and bilateral contracts' trading, the excess amount of electrical energy can be sold on a spot market for relatively low prices.

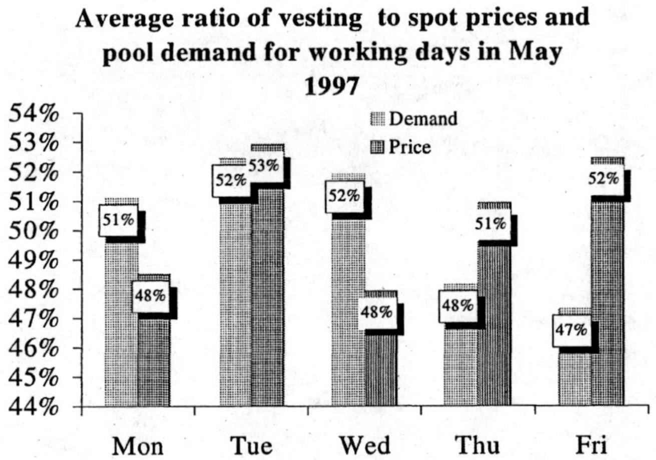


Fig. 3 Vesting contract demand and prices

## 4 BILATERAL HEDGE CONTRACTS

Independently on vesting contracts, market participants can enter voluntary, bilateral contracts in order to reduce the risk of price violations. Bilateral hedge contracts are typically constructed as:

- One Way Hedge contracts
- Two Way Hedge contracts
- Collar/Floor – Ceiling contracts

The aim of ***One Way Hedge*** contracts is to protect electricity Buyers against the increase of spot prices. A Seller and a Buyer agree to the level of a strike price above which a Seller rebates a Buyer costs of electricity purchase, see Fig. 4. A Buyer pays a Seller an annual fee to compensate the risk taken by a Seller. There are three major parameters of these contracts:

- strike price
- electricity quantity contracted
- annual fee

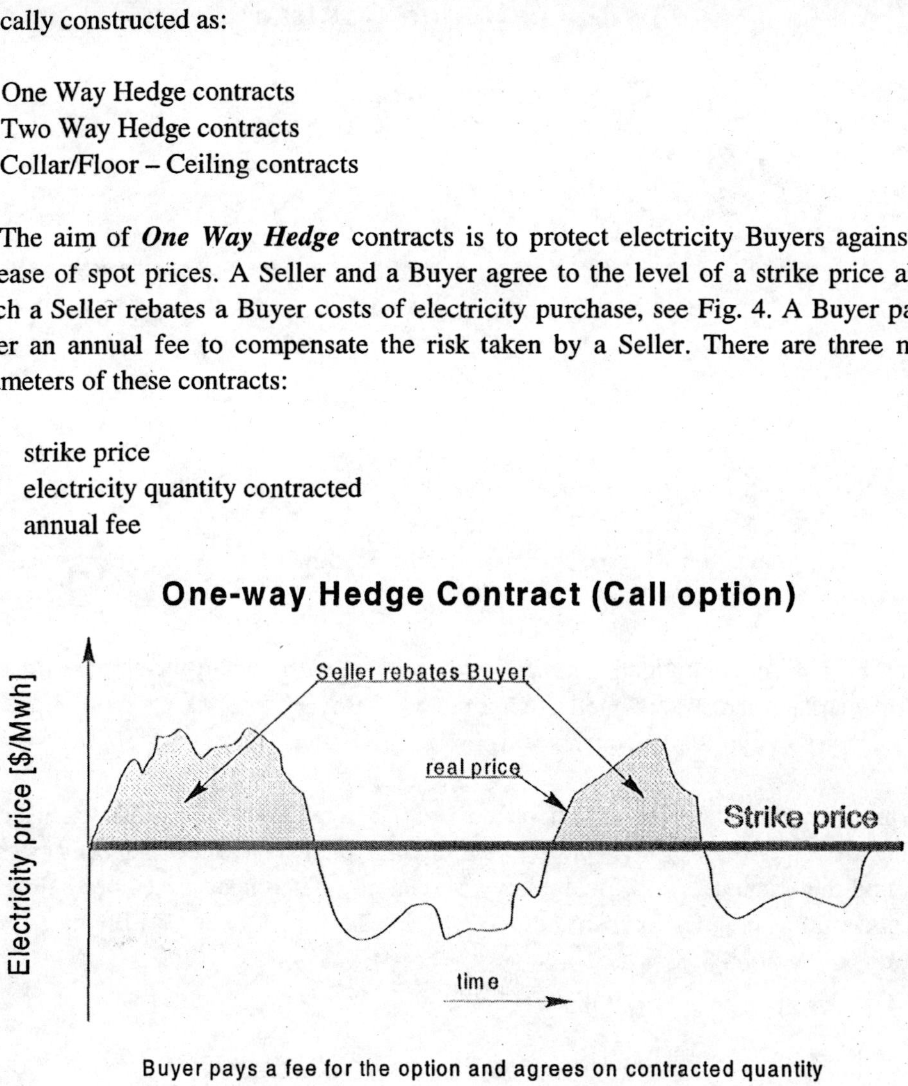


Fig. 4 One Way Hedge Contract (Call Option)

Characteristics of ***Two Way Hedge*** contract are presented in Fig. 5. A Seller and a Buyer agree to the value of strike price above which a Seller rebates a Buyer, however, when a spot price decreases below the level agreed, a Buyer rebates a Seller. This option gives a Seller and a Buyer a high level of protection against spot price variations. In practice, it means that electricity is purchased for a fixed price.

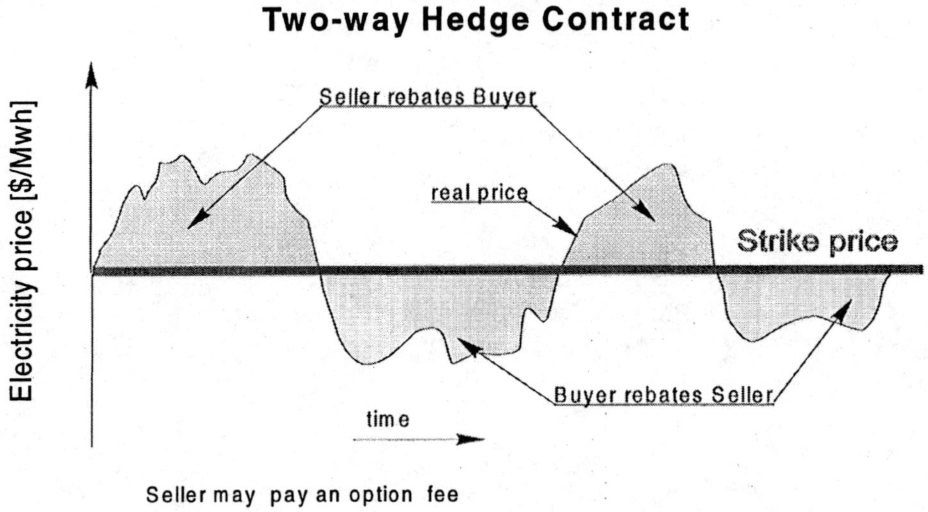


Fig. 5 Two Way Hedge Contract

Another hedge contract option is a ***Collar-type contract*** with two levels of strike prices:

- Upper strike price above which a Seller rebates a Buyer
- Lower strike price above which a Buyer rebates a Seller

When spot prices are above the upper strike price agreed, a Seller rebates a Buyer the differences resulting from high prices. The amount paid to a Buyer depends on spot prices and the amount of electrical energy purchased. When spot prices are below the lower strike price, a Seller compensates a Buyer. In addition, a Seller or a Buyer may pay an option fee, see Fig. 6.

## 5 Payments For Electrical Energy

Payments of electrical energy produced and consumed are carried out by the pool manager. However, compensation payments related to bilateral hedge contracts can also be transferred directly between Generators and Retailers. When market participants reveal the information on the amount of the energy contracted to pool managers (this is known as "netting out"), the pool managers include in electricity invoices only the difference between the energy measured and the amount of the energy contracted. On the other hand, when a hedge contract is not "netting out," Pool participants pay for the entire energy measured, compensating to each other differences resulting from higher or lower prices.

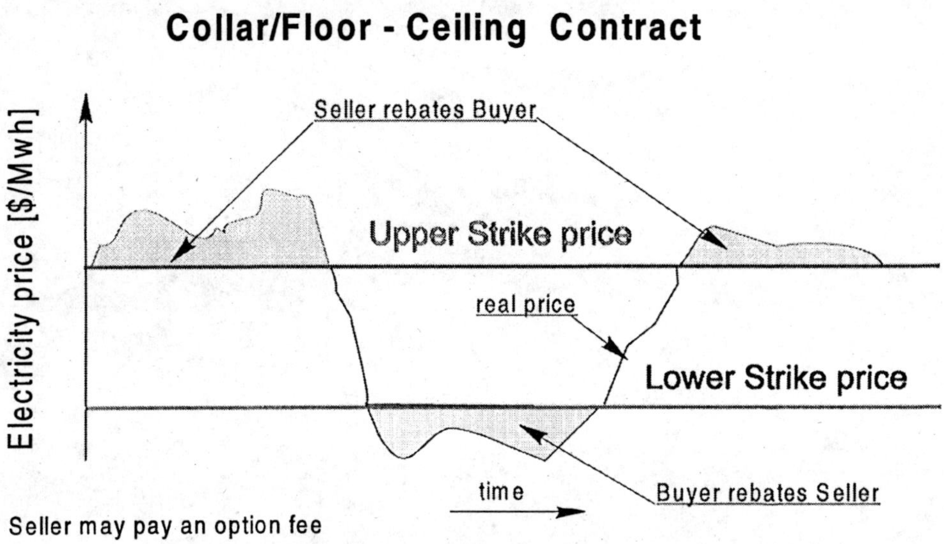


Fig. 6 Collar-type Hedge contract

## CASE STUDY – AN EXAMPLE OF PAYMENTS

Customer X, the chemical refinery operator, has a long term bilateral contract to purchase 80 MW of power at $50/MWh from Generator A who has a 200 MW generator. Customer's X plant has a demand of 110 MW. In order to reduce his exposure to the spot market, Customer X decides to buy an additional 10 MW of power from the market manager in the short-term forward contract at a market price of $40/MWh. The market manager purchases this power from other generators (see Fig. 7) [2].

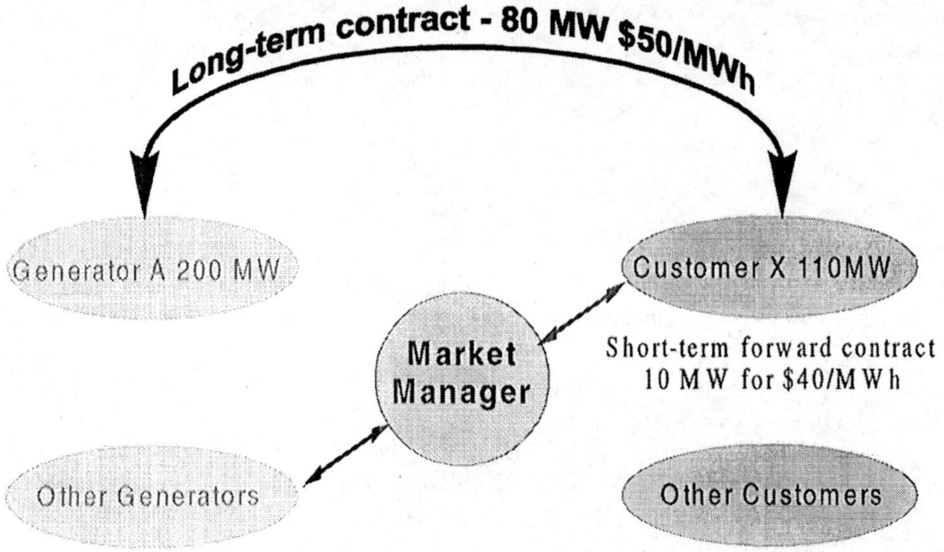


Fig. 7 Example of a long- and short-term contracts

Let us assume that for one hour of actual generation and consumption Customer X uses 110 MW, Generator A generates 150 MW, other Customers take a total of 540 MW, other Generators contribute to 500 MW, and the spot market price is $30/MWh. There are two possible methods of payments:

1. Customer X and Generator A decide not to notify the Pool manager on the contract quantity (not "netting out the contract"). Settlement flows are as follows (see Fig. 8). Customer X owes the market manager $3400 This amounts consists of $3000 for 100 MW at the spot price of $30/MWh and $400 for 10 MW at the price of $40/MWh, contracted in the short-term forward contract. Generator A is owned from the market manager $4500 for 150 MW supplied at the spot price. Under the terms of their long-term contract, Customer X owes Generator A $1600 representing the difference between the contract and spot price for the contact quantity of 80 MW.

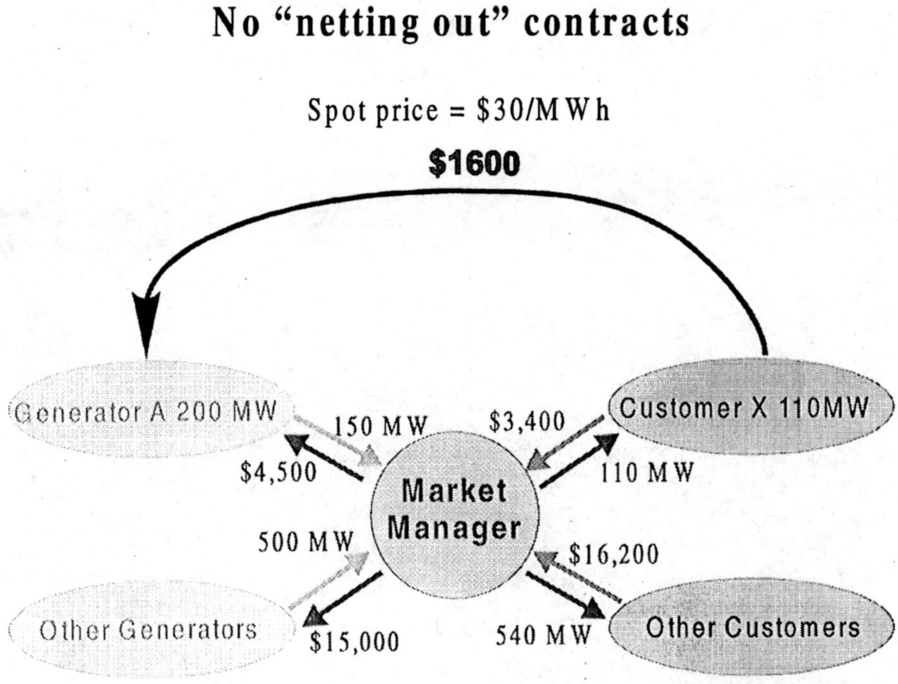


| | |
|---|---|
| Customer X owes market manager | 100 MW * $30/MWh = $3,000 for spot price<br>10 MW * $40/MWh = $ 400 for Short-term contract<br>Total = $3,400 |
| Generator A is owned from market manager | 150 MW * $30/MWh = $4,500 for spot sale |
| Customer X owes Generator A | 80 MW*($50 - $30)/MWh = $1,600 for Long-term contract |

Fig. 8 No "netting out" of a long-term contract

2. Customer X and Generator A have informed the market manager on their contract quantity of 80 MW ("netting out the contract"). Then, Customer X owns the market manager $1000: $600 for 20 MW at the spot price of $30/MWh and $400 for 10 MW at the price of $40/MWh contracted in the short-term forward contract. Generator A owns from the market manager $2100 for 70 MW at the spot price. Additionally, Customer X owes Generator A $4000 for the 80 MW supplied under the long-term contract (see Fig. 9).

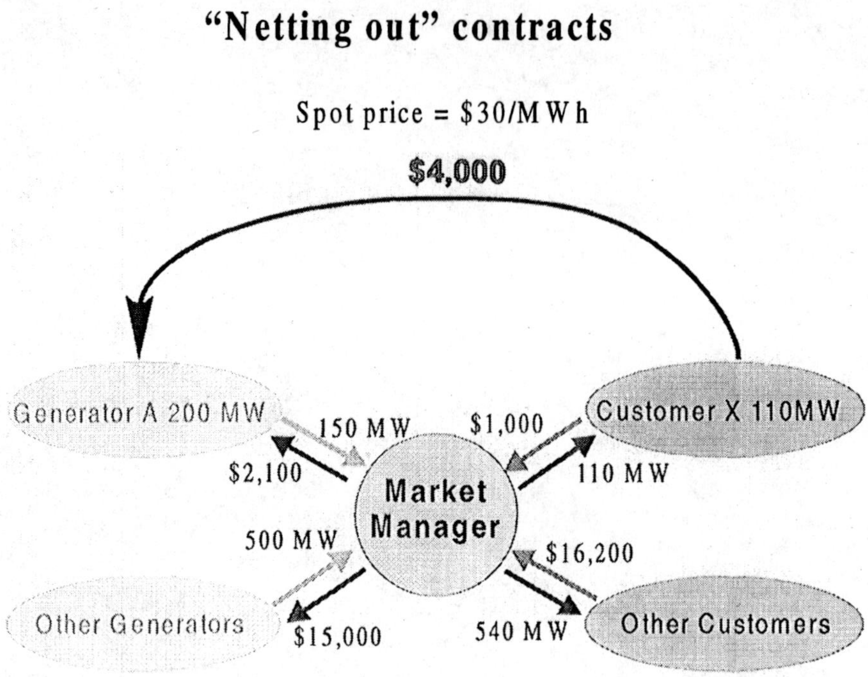


Customer X owes market manager — 20 MW * $30/MWh = $ 600 for spot price
10 MW * $40/MWh = $ 400 for Short-term contract
Total = $1,000

Generator A is owned from market manager 70 MW * $30/MWh = $2,100 for net traded at spot purchase

Customer X owes Generator A — 80 MW*$50 /MWh = $1,600 for Long-term contract

Fig. 9 Example of "netting out" of a long-term contract

Independently on whether or not the parties "netting out" their contract quantity from settlements, the total amount of payments is identical. In the above example, Customer X owes a total of $5000 for the 110 MW consumed over the hour.

## 6 SETTLEMENT PROCESS

The Office of the Regulator General has introduced regulations for settlements and metering. A central computerized system provides the settlement of the wholesale energy. Every participant has an energy account and a suitable meter that can register and store half-hourly electricity readings and can unload this data for settlements. The cash transfer takes place on specified number of days after the actual trading day. The settlement process in the Victorian market allows 28 days for the data collection, dispute, and the final settlements. Table 2 shows the timetable of the settlement procedure [2].

Table 2 Settlement time table

| Day | Action |
|---|---|
| -1 | Offers of energy and prices |
| 0 | Electricity trade |
| 1 ÷ 5 | Data collection and verification |
| 7 | Provisional settlement |
| 7 ÷ 8 | Data analysis and dispute |
| 18 ÷ 24 | Dispute Committee to resolve problems |
| 25 | Final settlement |
| 28 | Funds transfer |

## REFERENCES


1. The Electricity Supply Industry in Victoria. A Competitive Future – Electricity. Department of the Treasury, The Victorian Government, 1995.
2. VicPool Information Kit. Victorian Power Exchange, Victoria 1996.

****** 13 ******

# ELECTRICITY TARIFFS AND NETWORK CHARGES

**SOHAIL HASNIE**
Manager, Pricing, Office of the Regulator-General
Melbourne, Australia

**Summary:** This paper describes the broad regulatory framework for the regulation of electricity tariffs and network charges by Office of the Regulator-General (The Office). Until year 2001, when all electricity customers will be legible to select their choice of electricity retailer (or become non franchise), electricity prices for franchise customers are regulated by The Office. The Office also regulates charges for use of transmission and distribution networks. The Tariff Order provides a framework for the ORG activities in electricity price regulation. The selected electricity tariffs and network charges for 1997-1998 are presented in Appendix 1 and 2.

## 1 INTRODUCTION

Electricity tariffs and network charges are approved by the Office of the Regulator-General. Prices must be approved by the Office before customers can be charged using new prices. Office's role is 'Price Regulation' not "Price Surveillance" and broadly confined to:

- the areas of natural monopoly (the high voltage transmission network, the low voltage distribution system;
- VicPool or the wholesale market;
- the retail price for customers who do not have a choice of their electricity retailer – (franchise customers).

The key instrument that provides the details of price regulation is the Government's Victorian Electricity Supply Industry Tariff Order (the Tariff Order) effective from October 1994.

On 20 June 1995, the Governor in Council exercised the power in section 158A of the Electricity Industry Act 1993 to make an Order in Council, known as the Tariff Order, regulating prices through to 31 December 2000 and setting some parameters for the Office's to regulate prices after the year 2000.

The prices regulated by the Tariff Order [2] were locked in on privatisation and the Government's initial price regulatory regime cannot be varied other than by legislation requiring the approval of both Houses of the Parliament.

## 2 Regulation of Maximum Uniform Tariffs and Network Charges

### 2.1 Electricity Prices for Franchise Customers

Until customers are eligible to choose their retailer, they are known as franchise customers. Customers, who are eligible to choose a retailer, are known as non-franchise or contestable customers. Pending choice of a retailer being available, retail prices for franchise customers are regulated as announced by the Victorian Government in September 1994.

Prices for residential and rural customers were frozen to the middle of 1996, producing a real reduction of around 5%, for all but the largest power consumers. From July 1996 through to 31 December 2000, tariffs for all categories of smaller customers were to reduce in real terms applying a CPI-X formula.

These reductions in prices totalling 22% should be delivered to small and medium sized business customers December 2000. Real price reductions to residential, rural and farm consumers will be provided by way of a post July 1996 reduction of 2% (in real terms) and reductions thereafter of 1% per year (in real terms) until December 2000.

### 2.2 Tariff H Customers

Large industrial customers that were on tariff H prior to 3 October 1994 were given the choice of a safety net. Based on information provided to the government at time, tariff H prices were considered to be lower than the cost of service to those customers. Under the safety net arrangement, the customers had the choice of continuing with tariff H (adjusted to reflect changes in CPI) through to 31 December 2000, or to enter the contestable market. Once a Tariff H customer makes the decision to move into the contestable market, access to Tariff H is no longer available to that customer.

## 2.3 Other Charges

The Office also regulates, through to 31 December 2000, the charges, which may be imposed by:

- PowerNet and Victorian Power Exchange for connection to, and use of, the high voltage transmission grid; and
- the distribution companies for use of their low voltage distribution networks.

The Office is also required to approve, on a fair and reasonable basis, miscellaneous charges that may be imposed by the distribution companies, such as standard connection charges or charges for special metering which are not regulated directly by the Tariff Order. The Office is also responsible for the approval of the fees Victorian Power Exchange may charge participants in the wholesale market (VicPool) in respect of the operation of VicPool.

## 2.4 Prices for Non-Franchise Contestable Customers

While the Office continues to regulate the charge that a distribution company may impose on a non-franchise customer for delivering electricity across its network, it does not regulate the price that a retailer may charge a non-franchise customer for electricity - the price of electricity being subject to competition from other retailers. That is, non-franchise customers are free to shop around to get the best price for electricity.

In summary, a non-franchise customer's purchase consists of two distinct components:

- a delivery or transport component, which is charged at a regulated rate; and
- an energy component, which is charged at competitive rates.

This gives a non-franchise customer the opportunity to negotiate competitive prices and terms for electricity, secure in the knowledge that its local distribution company must deliver it at a regulated tariff [1].

## 3 Components of Electricity Prices

The price which a retailer charges a customer for electricity (as distinct from other charges which a distribution company may impose, eg - for excluded services) reflects the following:

- energy costs;
- pool fees;
- network costs; and
- marketing costs and a profit margin.

## 3.1 Energy Costs

Electricity costs reflect the net price paid by a retailer for electricity acquired through the pool at pool prices taking into account the effect of any hedging contracts the retailer may have entered to guard against fluctuations in the pool prices.

The pool prices and hedging contracts are not regulated by the Tariff Order nor by the Office.

## 3.2 Pool Fees

Pool fees are the fees payable by a retailer to Victorian Power Exchange in accordance with the Pool Rules. Pool fees are determined in accordance with and are regulated by the Tariff Order. Pool fees paid by a retailer reflect the following cost elements:

- pool costs, or Victorian Power Exchange's expenses of operating and administering the pool; and
- ancillary services charges which are the charges payable by Victorian Power Exchange to generators for ancillary services required for secure system operations. Ancillary services include such services as black start capability and automatic generation control capability.

## 3.3 Network Tariffs

Network tariffs include the cost of using the high voltage transmission grid and the distribution network to deliver electricity to the retailer's customers. Network tariffs are regulated by the Tariff Order until December 2000. Thereafter, distribution network tariffs will be regulated by the Office and transmission network tariffs by ACCC (Australian Competition and Consumer Commission).

The retailer pays (whether it is the retail arm of the distribution company or an independent retailer) the local distribution company a network tariff (approved by the Office as complying with the requirements of the Tariff Order). The details of the connection agreement between the independent retailer and the host DB is governed by the use of system agreement between the retailer and the local distribution company.

Where the retailer is the retail arm of the local distribution company, the retailer's charges to the non-franchise customer will be an amount equivalent to the network tariff the local distribution company would charge another retailer for use of its system.

### 3.4 Other Elements

Other elements of a retailer's charges to a non-franchise customer include the retailer's own marketing costs and profit margins. These other elements of a retailer's charges are not regulated by the Tariff Order or by the Office.

### 3.5 Local Connection Charges

Local connection charges may be applied by a local distribution company to its customers for connection to its distribution network. A local distribution company's connection charges are not regulated by the Tariff Order but are regulated by the Office subject to a fair and reasonable test. The process, by which the Office approves such connection and other charges that may be imposed by a distribution company, is canvassed in the Office's Electricity Industry Guideline No. 1 [3].

## 4 Pool Charges to Non Franchise Customers

Non franchise customers have a choice. They can purchase electrical energy from the retail arm of the distribution company, any independent electricity retailer, or directly from the pool. When a non franchise customer buys electricity from VicPool, the payments are:

- for the electricity at the prevailing Pool prices which is determined as result of competing generators at the wholesale market
- pool fees, as specified in the Tariffs Order,
- network costs as per the published network tariffs payable by the non franchise customer under its use of system agreement with its local distribution company.

## 5 The Government's Initial Prices

In exercising its power to regulate initial prices by way of the Tariff Order the Government has:

- set CPI for maximum uniform tariffs (varying in accordance with CPI based formulas) for franchise customers through to 31 December 2000;
- set a CPI-X revenue cap through to 31 December 2000 for the transmission business (including the network charge payable by Victorian Power Exchange and connection charges);
- empowered the Office to each year approve Victorian Power Exchange's pool fees and use of system fees on the basis of cost recovery; and

- set constraints on the network tariffs charged by the distribution companies' wire businesses through to 31 December 2000.

## 6 Reducing Cost of Electricity

### 6.1 Residential and Rural Customers

The maximum prices, which a distribution company may charge residential and rural customers, were frozen from 3 October 1994 to 30 June 1996. Since 1 July 1996 there will be a 2% real reduction in those maximum prices, followed by a 1% real reduction each year thereafter until the year 2000 when franchise customers move to choice in a fully competitive market.

Rural and farm customers also have maximum uniform tariffs during the franchise period and prices held at the level comparable to tariffs for metropolitan residential customers.

Residential and rural tariff structures will be maintained until 31 December 2000. The tariffs within those structures have been set by the Government as the Maximum Uniform Tariffs (or MUTs). The summary of the price reduction is as follows:

- no change to 1 July 1996, which equates to an expected 6.4%;
- real reduction based on inflation;
- thereafter maximum uniform tariffs are adjusted on the following basis:

  - 1 July 1996 CPI – 2% real reduction
  - 1 July 1997 CPI – 1% real reduction
  - 1 July 1998 CPI – 1% real reduction
  - 1 July 1999 CPI – 1% real reduction
  - 1 July 2000 CPI – 1% real reduction
  - December 2000 – customer choice.

### 6.2 Small and Medium Businesses

Small and medium businesses will receive at least real electricity reductions totalling 22% before year 2000. The reduction will be achieved by maintaining the small and medium businesses tariff structure existing at 3 October 1994 until business may choose between retailers. Those tariff structures are indexed on average as follows

- 1 July 1995 CPI – 10% reduction
- 1 July 1996 CPI – 5% reduction
- 1 July 1997 CPI – 5% reduction

- 1 July 1998 CPI – 1% reduction
- 1 July 1999 CPI – 1% reduction
- 1 July 2000 CPI – 1% reduction
- December 2000 – customer choice

### 6.3 Large Business Customers

Large business customers which are franchise customers, if they select to pay for electricity due to the maximum tariffs, will generally be on either tariffs H, L, V or E5 – Appendix 1. These tariffs are subject to an annual CPI cap through to December 2000, which means that they cannot increase in real terms during the term of the Tariff Order.

## 7 Network Charges and Pool Fees

### 7.1 Distribution Network Tariffs

Network tariffs are charged by a distribution company to recover:

- the transmission use of system (TUOS) charges;
- the transmission company's (PowerNet) connection charges;
- the distribution company's distribution use of system (DUOS) costs.

To minimise the impact on customer categories or classes of any rebalancing (or shock rate), the Tariff Order imposes a CPI + 2% limit on the increase in the average price of each Network Tariff that may be applied in any year to December 2000. For details of the CPI+2 constrains on re-balancing of individual network tariffs please refer to the formulas in the tariff order a copy of which is available at www.reggen.vic.gov.au.

### 7.2 Transmission Network Charges

The transmission company, (Power Net Victoria) charges the pool manager (Victorian Power Exchange) a network charge for use of the transmission system. The network charge is determined by reference to Power Net Victoria's operating, maintenance, administration and management costs together with a standardised capital charge.

Power Net Victoria also directly charges generators and the distribution businesses for connection to the high voltage transmission network.

The Tariff Order applies a CPI-X revenue cap to Power Net Victoria's revenue derived from its network charges to December 2000.

The pool manager passes the transmission network charges to the generators and the distribution companies. These charges, along with Victorian Power Exchange's own costs of operating the transmission system, create transmission use of system (TUOS) charges.

### 7.3 Pool Fees

Pool fees reflect the costs incurred by Victorian Power Exchange in operating the wholesale electricity market and the charges it incurs for ancillary services provided by the generators. Pool fees are set in accordance with the Pool Rules and the Tariff Order. The pool fees must be approved by the Office of the Regulator-General.

## 8 Procedure for Submitting and Approving

### 8.1 Procedures for Submitting

A licensee may submit a statement of charges for approval by the Office at any time. If a licensee has had a statement of charges approved by the Office, it must submit any proposed variations to that statement for approval before implementation the variation. A statement of charges must be submitted within 20 business days on a written request of the Office.

In some cases when consideration of the issue involves consultation with third parties, the Office will inform the licensee of the time frames in which consultations to be completed.

### 8.2 The Office's Criteria for Approving

In deciding whether to approve a statement of charges the Office has regard to its objectives in the Office of the Regulator-General Act 1994 and the Electricity Industry Act 1993.

While the Office would expect to see differences in charges between different customers or classes of customers attributable to the different:

- capacity required to provide service;
- volume or quantity in which goods or services are supplied;
- places to or from which the services are to be supplied;
- periods for which the services are to be supplied; or
- performance characteristics at which the services are to be supplied,

however, it will examine any cross-subsidisation between a licensee's:

- regulated and non-regulated activities;
- franchise and non franchise customers,

to ascertain whether it is anti –competitive or otherwise not in the interest of customers.

## 9 Retail Tariffs for Franchise Customers in 1997-98

The selected retail tariffs for franchise customers are presented in Appendix 1. The main categories of these tariffs are as follows:

- Residential
- General purpose low voltage
- Community Service
- Farm
- General purpose high voltage
- Public lighting
- Miscellaneous charges

Some tariffs for example Off-peak Storage Water Heating Tariff Y8, are available only to installations currently taking supply under this tariffs. This tariffs were 'closed tariffs' at the time of de-regulation and aims to allow customers to move gradually to new proposed tariffs (in this case Y6).

## 10 Network Tariffs In 1997-98

The selected network charges are presented in Appendix 2. The charges are different for five distribution businesses. The difference reflects the specific supply conditions of five distributors.

The main categories of network tariffs include:

- Low voltage tariffs for nominal voltage smaller than 1000V
- High voltage tariffs for nominal voltage between 1000V and 22,000V
- Sub transmission tariffs for voltage equal to 22,000V and above.

The low voltage tariffs are designed to reflect size of customers and end energy consumed. They include:

- Small single and two rate tariffs
- Dedicated circuit tariffs
- Medium single and two rate tariffs
- Large tariff

- Large low voltage up to 750MWh per annum
- Large two rate tariff
- Large tariff A and C

Two-rate tariffs and tariffs for large customers are constructed with peak and off-peak rates.

## REFERENCE

1. Electricity Industry Regulatory Statement. The Office of the Regulator-General.
2. Electricity Industry Act 1994. Section 158A. Victorian Electricity Supply Industry Tariff Order. Melbourne, 1995
3. Electricity Industry Guideline No1. "Access to, and use of, distribution systems statements of approved charges and charges for other services". The Office of the Regulator-General. Melbourne 1995.

# APPENDIX 1

# SELECTED RETAIL TARIFFS FOR FRANCHISE CUSTOMERS FOR 1997/98

## 1 RESIDENTIAL

***1.1 Winner Tariff GH/GL***

| | |
|---|---|
| Energy Prices: | |
| Peak Periods (7 am to 11 pm Mon to Fri) | |
| All consumption | 17.55 c/kWh |
| Off-Peak Periods (All other times) | |
| All consumption | 3.79 c/kWh |

***1.2 Residential Tariffs GD and GR***

| | |
|---|---|
| First 1020 kWh/quarter | 12.03 c/kWh |
| Balance | 12.68 c/kWh |

***1.3 Off-Peak Load Managed Storage Water Heating Tariff Y6/YT***

| | |
|---|---|
| All consumption | 3.79 c/kWh |

Supply under Tariff Y6/YT is available for 6 hours nightly for permanently wired water storage heaters of approved types meeting load management requirements.

***1.4 Off-Peak Storage Water Heating Tariff Y8***

(Available only to installations currently taking supply under this tariff)

| | |
|---|---|
| All consumption | 4.52 c/kWh |

Supply under Tariff Y8 is available for 8 hours nightly for permanently wired storage water heaters of approved types.

***1.5 Off-Peak Storage Space Heating Tariff J6/JT***

| | |
|---|---|
| All consumption | 3.79 c/kWh |

Supply under Tariff J6/JT is available for 7 hours nightly plus 3 hours each afternoon for permanently wired storage space heaters of approved types where supply is taken in conjunction with Tariff Y6/YT

***1.6 Off-Peak Storage Space Heating Tariff J8***

| | |
|---|---|
| All consumption | 4.52 c/kWh |

Supply under Tariff J8 is available for 7 hours nightly plus 3 hours each afternoon for permanently wired storage space heaters of approved types, where supply is taken in conjunction with Tariff Y8.

***1.7 Off-Peak Storage Space Heating Tariff J***

| | |
|---|---|
| All consumption | 4.52 c/kWh |

Supply under Tariff J is available for 7 hours nightly plus 3 hours each afternoon for permanently wired storage space heaters of approved types, where supply is not taken in conjunction with a storage water heating Tariff.

***1.8 Supply Charge***

| | |
|---|---|
| On each account rendered | $34.41 /qtr |

## 2 General Purpose Low Voltage

2.1

***(a) General Purpose Tariff E***

| | |
|---|---|
| First 7000 kWh/month | 16.56 c/kWh |
| Balance | 12.61 c/kWh |
| Supply Charge –on each account rendered | $12.89 /mth |

***(b) Tariff E1***

(Available only to installations currently taking supply under this tariff).

| | |
|---|---|
| Energy Prices: | |
| 7 am to 11 pm | |
| First 5500 kWh/month | 19.29 c/kWh |
| Balance | 12.55 c/kWh |
| 11 pm to 7 am | |
| All consumption | 4.03 c/kWh |

Supply Charge –on each account rendered $12.89 /mth

***2.2 Commercial Residential Tariff G***

(Available only to installations currently taking supply under this tariff)

| | |
|---|---|
| First 5000 kWh/month | 16.99 c/kWh |
| Balance | 13.57 c/kWh |
| Supply Charge –on each account rendered | $12.89 /mth |

***2.3 Metered Demand Tariff V***

(Available only to installations currently taking supply under this tariff)

| | |
|---|---|
| Energy Price | 5.29 c/kWh |
| Demand Price | $18.45 |

/kW/mth

The minimum chargeable demand is the higher of 100 kW or 60% of agreed demand.

***2.4 Contract Demand Time-of-Use Tariff L***

| | |
|---|---|
| Energy Prices | |
| Peak Periods (7 am to 11 pm Monday to Friday) | |
| All consumption | 11.09 c/kWh |
| Off-Peak Periods (All other times) | |
| All consumption | 2.27 c/kWh |
| Demand Price | $3.68 /kW/mth |

Minimum chargeable demand 250 kW

Supply Charge -oneach account rendered $345.03 /mth

***2.5 General Purpose Time-of-Use Tariff D***

| | |
|---|---|
| Energy Prices | |
| Peak Periods (7 am to 11 pm Monday to Friday) | |
| All consumption | 16.74 c/kWh |
| Off-Peak Periods (All other times) | |
| All consumption | 3.30 c/kWh |
| Supply Charge –on each account rendered | $12.89 /mth |

***2.6 Resale Price***

Price for reselling (applicable to premises such as caravan parks where the proprietor is authorised under the Electric Light and Power Act to charge customers for electricity actually used).

Maximum resale price where token or prepayment meters are installed:

all energy 12.68 c/kWh

Resale price where kilowatt hour (kWh) meters
are installed: as per Tariff GD/GR.

## 3 Community Service

***3.1 Community Service Tariff N***

All consumption 14.61 c/kWh

Supply Charge –on each account rendered $14.81 /mth

***3.2 Tariff N1***

(Available only to installations currently taking supply under this tariff)

Energy Prices

7am to 11pm

All consumption 14.61 c/kWh

11pm to 7am

All consumption 4.63 c/kWh

Supply Charge –on each account rendered $14.81 /mth

## 4 Farm

***4.1 Farm Tariff B***

Energy Prices:

First 690 kWh/quarter 20.12 c/kWh

Next 14310 kWh/quarter 18.59 c/kWh

Next 72000 kWh/quarter 15.54 c/kWh

Balance 14.41 c/kWh

Supply Charge –on each account rendered $44.41 /qtr

***4.2 Tariff B1***

(Available only to installations currently taking supply under this tariff).

Energy Prices:

7 am to 11 pm

First 690 kWh/quarter 20.12 c/kWh

Next 14310 kWh/quarter 18.59 c/kWh

Next 72000 kWh/quarter 15.54 c/kWh

Balance 14.41 c/kWh

11 pm to 7 am -

All consumption 4.63 c/kWh

Supply Charge –on each account rendered $44.41 /qtr

## 5 General Purpose High Voltage

### *5.1 General Purpose Tariff E5*

(Available only to installations currently taking supply under this tariff).

Energy Prices:

7 am to 11 pm

First 5500 kWh/month 22.18 c/kWh

Balance 14.45 c/kWh

11 pm to 7 am

All consumption 4.65 c/kWh

Supply Charge –on each account rendered $15.77 /mth

### *5.2 Contract Demand Time-of-Use Tariff H*

| | OPTION 1 | OPTION 2 | OPTION 3* |
|---|---|---|---|
| Voltage levels (kV) | 6.6/11/22 | 66 | 220 & above |
| Energy Prices (c/kWh) | | | |
| Peak Periods (7am - 11pm Mon to Fri) | 9.14 | 7.39 | 6.78 |
| Off-Peak Periods (All other times) | 1.93 | 1.78 | 1.70 |
| Demand Price ($/kW/month) | 3.78 | 4.82 | 4.90 |
| Minimum Chargeable demand (kW) | 1000 | 10,000 | 40,000 |

*This option is not offered by Eastern Energy Limited

## 6 Public Lighting

| Lanterns on Current Offer TYPE AND RATING OF | | 24 HOUR ALL NIGHT $ pa | DAILY $ pa |
|---|---|---|---|
| Mercury | 80W | 80.67 | - |
| Mercury | 125W | 105.86 | - |
| Sodium | 100W | 98.64 | - |
| Sodium | 150W | 151.12 | 333.63 |
| Sodium | 250W | 81.05 | 423.07 |
| Sodium | 400W | 279.97 | - |

## APPENDIX 2

# NETWORK TARIFFS IN 1997-98

LOW VOLTAGE TARIFFS (NOMINAL VOLTAGE < 1000 VOLTS)

| | units | Eastern | Solaris | CitiPower | United | Powercor |
|---|---|---|---|---|---|---|
| Small single rate tariff | | | | | | |
| -Standing charge | $pa | 63.000 | 53.000 | 46.04 | 51.233 | 48.784 |
| -Unit rate | c/kWh | 6.281 | 5.817 | 5.486 | 5.678 | 7.253 |
| Small single rate - bulk | | | | | | |
| -Standing charges | $pa - | - | 32.36 | - | -- | |
| Unit rate | c/kWh | - | - | 4.543 | - | - |
| Smallt wo rate tariff | | | | | | |
| -Standing charge | $pa | 100.000 | 87.800 | 84.57 | 80.535 | 88.732 |
| -Peak unit rate | c/kWh | 7.989 | 7.320 | 7.088 | 7.120 | 9.599 |
| -Off peak unit rate | c/kWh | 1.362 | 0.934 | 1.220 | 1.289 | 1.326 |
| Small two rate - bulk | | | | | | |
| -Standing charge | $pa | - | - | 71.89 | - | -- |
| Peak unit rate | c/kWh | - | - | 6.037 | - | -- |
| Off peak unit rate | c/kWh | - | - | 0.979 | - | - |
| Dedicated circuit tariff | | | | | | |
| -Standing charge | $pa | 39.000 | 62.000 | 62.78 | 52.173 | 50.121 |
| -Off peak unit rate | c/kWh | 1.362 | 0.934 | 1.223 | 1.106 | 1.306 |
| Dedicated supplies - bulk | | | | | | |
| -Standing charge | $pa | - | - | 56.51 | - | -- |
| Off peak unit rate | c/kWh | - | - | 0.979 | - | - |
| Medium single rate tariff | | | | | | |
| -Standing charge | $pa | 100.000 | 76.300 | 52.49 | 70.576 | 66.650 |

| | | | | | | |
|---|---|---|---|---|---|---|
| -Unit rate | c/kWh | 6.889 | 6.418 | 5.967 | 6.866 | 7.914 |
| **Medium single rate tariff - bulk** | | | | | | |
| -Standing charge | $pa | - | - | 43.27 | - | -- |
| Unit rate | c/kWh | - | - | 4.982 | - | - |
| **Medium two rate tariff - 5 day** | | | | | | |
| -Standing charge | $pa | 221.000 | 202.400 | 198.52 | 190.411 | 180.05 |
| -Peak unit rate | c/kWh | 7.850 | 7.502 | 6.90 | 7.174 | 9.182 |
| -Off peak unit rate | c/kWh | 1.945 | 1.077 | 1.392 | 2.149 | 1.760 |
| **Medium two rate - 5 day - bulk** | | | | | | |
| -Standing charge | $pa | - | - | 185.89 | - | -- |
| Peak unit rate | c/kWh | - | - | 5.566 | - | -- |
| Off peak unit rate | c/kWh | - | - | 1.052 | - | - |

| | units | Eastern | Solaris | CitiPower | United | Powercor |
|---|---|---|---|---|---|---|
| Medium two rate tariff - 7 day | | | | | | |
| -Standing charge | $pa | 221.000 | 202.400 | 179.01 | 190.411 | 180.05 |
| -Peak unit rate | c/kWh | 7.223 | 6.483 | 6.006 | 6.918 | 8.357 |
| -Off peak unit rate | c/kWh | 1.976 | 1.077 | 1.427 | 2.255 | 1.760 |
| Medium two rate tariff - 7 day - bulk | | | | | | |
| -Standing charge | $pa | - | - | 166.58 | - | -- |
| Peak unit rate | c/kWh | - | - | 4.866 | - | -- |
| Off peak unit rate | c/kWh | - | - | 1.086 | - | - |
| Unmetered supplies | | | | | | |
| -Peak unit rate | c/kWh | 7.850 | 6.479 | 6.995 | 7.827 | 9.128 |
| -Off peak unit rate | c/kWh | 1.945 | 1.077 | 1.386 | 1.004 | 1.953 |
| Large Tariff | | | | | | |
| -Standing charge | $pa | 3,070 | 3,702.8 | 3,933.08 | 3,255 | 2,984 |
| -Peak unit rate | c/kWh | 1.926 | 2.160 | 1.674 | 1.484 | 2.551 |
| -Off peak unit rate | c/kWh | 1.337 | 0.406 | 0.757 | 0.693 | 1.103 |
| -Demand | $/kW pa | 78.450 | 55.173 | 59.64 | 59.374 | 84.059 |
| Minimum chargeable demand kW | 120 | 250 | 120 | 120 | 250 | |
| Large low voltage up to 750 MWh | | | | | | |
| -Standing charge | $pa | 3,070 | - | - | - | -- |
| Peak unit rate | c/kWh | 1.926 | - | - | - | -- |
| Off peak unit rate | c/kWh | 1.337 | - | - | - | -- |
| Demand | $/kVA pa | 62.760 | - | - | - | - |
| Minimum chargeable demand 150 kVA | | | | | | |
| Large two rate tariff - 7 day peak* | | | | | | |
| -Standing charge | $pa | - | - | 171.55 | - | -- |
| Peak unit rate | c/kWh | - | - | 5.436 | - | -- |
| Off peak unit rate | c/kWh | - | - | 1.331 | - | -* |
| This tariff is restricted to customers previously taking supply under retail tariff E1 or N1 and who are consuming more than 400 MWh pa - new in 1996/97 | | | | | | |
| Large Tariff A | | | | | | |
| -Standing charge | $pa | - | - | - | - | 2,864.3 |
| -Peak unit rate | c/kWh | - | - | - | - | 2.414 |
| -Off peak unit rate | c/kWh | - | - | - | - | 1.059 |
| -Demand | $/kW pa | - | - | - | - | 80.679 |

| | | | | | | |
|---|---|---|---|---|---|---|
| Minimum chargeable demand 250 kW | | | | | | |
| Large Tariff C# | | | | | | |
| -Standing charge | $pa | - | - | - | - | 2,924.0 |
| -Peak unit rate | c/kWh | - | - | - | - | 2.502 |
| -Off peak unit rate | c/kWh | - | - | - | - | 1.081 |
| -Demand | $/kW pa | - | - | - | - | 82.360 |
| Minimum chargeable demand 250 kW | | | | | | |

HIGH VOLTAGE TARIFFS (NOMINAL VOLTAGE >=1000 VOLTS AND <=22,000 VOLTS)

| | units | Eastern | Solaris | CitiPower | United | Powercor |
|---|---|---|---|---|---|---|
| *High voltage* | | | | | | |
| -Standing charge | $pa | 3,210 | 2,174 | 2,646.16 | 1,820 | 1,691.1 |
| -Peak unit rate | c/kWh | 1.430 | 1.695 | 1.573 | 1.069 | 1.920 |
| -Off peak unit rate | c/kWh | 0.426 | 0.126 | 0.548 | 0.486 | 0.343 |
| -Demand | $/kW pa | 62.250 | 44.020 | 46.35 | 39.662 | 83.064 |
| Minimum chargeable demand 1,000 kW | | | | | | |
| | | | | | | |
| High voltage A | | | | | | |
| -Standing charge | $pa | - | - | - | - | 1,691.1 |
| -Peak unit rate | c/kWh | - | - | - | - | 1.701 |
| -Off peak unit rate | c/kWh | - | - | - | - | 0.244 |
| -Demand | $/kW pa | - | - | - | - | 62.300 |
| Minimum chargeable demand 1,000 kW | | | | | | |
| | | | | | | |
| High voltage C | | | | | | |
| -Standing charge | $pa | - | - | - | - | 1,656.9 |
| -Peak unit rate | c/kWh | - | - | - | - | 1.874 |
| -Off peak unit rate | c/kWh | - | - | - | - | 0.336 |
| -Demand | $/kW pa | - | - | - | - | 81.385 |
| Minimum chargeable demand 1,000 kW | | | | | | |
| | | | | | | |
| SUBTRANSMISSION TARIFFS | | | | | | |
| ***Subtransmission Tariff (Nominal voltage =22,000 volts)*** | | | | | | |
| -Standing charge | $pa | - | - | 2,536.57 | 1,820 | -- |
| Peak unit rate | c/kWh | - | - | 1.862 | 0.310 | -- |
| Off peak unit rate | c/kWh | - | - | 0.136 | 0.210 | -- |
| Demand | $/kW pa | - | - | 11.89 | 8.678 | - |
| Minimum chargeable demand 10,000 kW | | | | | | |
| | | | | | | |
| ***Subtransmission Tariff (Nominal voltage >22,000 volts)*** | | | | | | |
| -Standing charge | $pa | 17,885 | 20,093.8 | - | - | 21,739 |
| -Peak unit rate | c/kWh | 0.980 | 0.775 | - | - | 1.315 |
| -Off peak unit rate | c/kWh | 0.027 | 0.013 | - | - | 0.028 |
| -Demand $/kW | pa | 8.730 | 10.556 | - | - | 10.973 |
| Minimum chargeable demand 10,000 kW | | | | | | |
| | | | | | | |
| Subtransmission Tariff (Nominal voltage >22,000 volts)A | | | | | | |
| -Standing charge | $pa | - | - | - | - | 21,109 |
| -Peak unit rate | c/kWh | - | - | - | - | 1.327 |
| -Off peak unit rate | c/kWh | - | - | - | - | 0.027 |
| -Demand | $/kW pa | - | - | - | - | 10.655 |

| | | | | | | |
|---|---|---|---|---|---|---|
| Minimum chargeable demand 10,000 kW | | | | | | |
| | | | | | | |
| Subtransmission Tariff (Nominal voltage >22,000 volts)G | | | | | | |
| -Standing charge | $pa | - | - | - | - | 21,109 |
| -Peak unit rate | c/kWh | - | - | - | - | 1.328 |
| -Off peak unit rate | c/kWh | - | - | - | - | 0.027 |
| -Demand | $/kW pa | - | - | - | - | 10.655 |
| Minimum chargeable demand 10,000 kW | | | | | | |
| | | | | | | |
| Subtransmission Tariff (Nominal voltage >22,000 volts)S | | | | | | |
| -Standing charge | $pa | - | - | - | - | 20,890 |
| -Peak unit rate | c/kWh | - | - | - | - | 1.336 |
| -Off peak unit rate | c/kWh | - | - | - | - | 0.026 |
| -Demand | $/kW pa | - | - | - | - | 10.544 |
| Minimum chargeable demand 10,000 kW | | | | | | |

****** 14 ******

# DEMAND AND SPOT PRICE FOR ELECTRICAL ENERGY


WLADYSLAW MIELCZARSKI AND GRAZYNA MICHALIK

Monash University, Department of Electrical and Computer Systems Engineering

Melbourne, Australia, Email: Grazyna@basil.eng.monash.edu.au



**Summary:** The paper discusses demand for electricity and prices in vesting and bilateral hedge contracts. The detailed analysis shows trends of spot prices in the Victorian market (Vicpools) and the National Electricity Market.


## 1 INTRODUCTION

Victoria was the first Australian State that introduced the open electricity market. This wholesale market, called VicPool started to operate in July 1994 and was transformed into National Electricity Market (NEM) in May 1997. Currently, NEM consists of two regions: Region 1 comprising Victoria and South Australia, and Region 2 that includes New South Wales and Australian Capital Territory. In addition, Snowy Mountains Hydro Generation is participating in the market. However, due to transmission link constrains, the energy transfer between regions is limited.

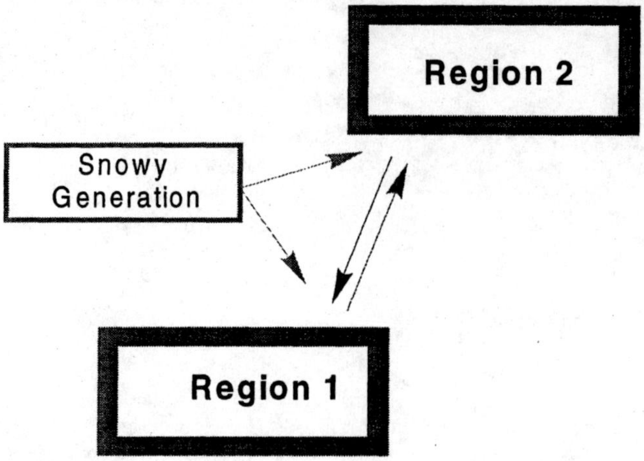


Fig. 1 National Electricity Market in Australia in 1997

There are three levels of electricity wholesale trade in an open electricity market in Victoria:

- Vesting contracts
- Bilateral contracts (long and short term)
- Spot trading

## 2 Vesting Contracts

Vesting contracts are financial instruments arranged between power stations and energy retailers by the Victorian Government before the introduction of the open electricity market. They aim at the market stabilization allowing power stations to recover a long-term marginal cost. Vesting contracts will last to year 2000.

Vesting contracts split energy consumption during a year in three characteristic seasons and two distinct categories of days. The seasons for electricity demand were assigned as:

- Cold season – from 1 June to 31 August
- Mild season – from 1 September to 30 November and from 1 April to 31 May
- Hot Season – from 1 December to 31 March

There are tow characteristic days for electricity demand:

- Working days from Monday to Friday
- Non working days include: Saturdays, Sundays, National and State Holidays, and a period from 21 December to 19 January.

Figure 2 demonstrates electricity demand covered by vesting contracts in the beginning of the competitive electricity market in Victoria. Tuesday, in the first week of July has been selected to show the demand cover by vesting contracts. As seen the vesting contracts of five distribution businesses (DB1 – DB5), which entered the competitive market on 1 July 1995, comprise nearly 4000MW of electricity demand. Patterns of vesting contracts reflect patterns of energy consumption in Victoria.

Amount of electrical demand covered by vesting contracts is gradually reduced when the competitive electricity markets is developing and more customers become non-franchise customers eg. they have right to select an electricity supplier. The electricity demand covered by vesting contract in July 2000 is shown in Figure 3. Half a year before, a fully competitive markets is introduced for all customers, vesting contracts cover less that 2000 MW of electricity demand. Vesting contracts expire in January 2000.

Vesting contracts as a percentage of the total demand are shown in Figure 4 and 5. While in the start of the competitive markets in Victoria, vesting contracts covered almost 80% of the total demand in this time, in July 2000, vesting contracts will cover less that 35% of the total demand from July 1995. The total demand in July 200 should be higher than in July 1995 as demand of electricity in Victoria increases 3-5% every year so the cover of electricity demand by vesting contracts in July200 should be less that 30% of the total demand.

Prices of electrical energy in vesting contracts depend on a season and a day of the week. Vesting contract prices also reflect patterns of energy use and patterns of spot prices. Examples of vesting contract prices are shown in Figure 6 and 7 for a working (Tuesday) and non working day (Sunday). Vesting contract electricity prices in working days are high reaching over $60/MWh in peak demand. Electricity prices on Sundays are lower; however they are very high compering to spot prices with the same periods. Electricity prices in vesting contract are gradually reduced towards year 2000.

High electricity prices in vesting contracts allow power stations to recover their fixed costs of generation encouraging the trade of electrical energy excess on the spot market for prices close to variable costs. This drives spot prices to very low levels.

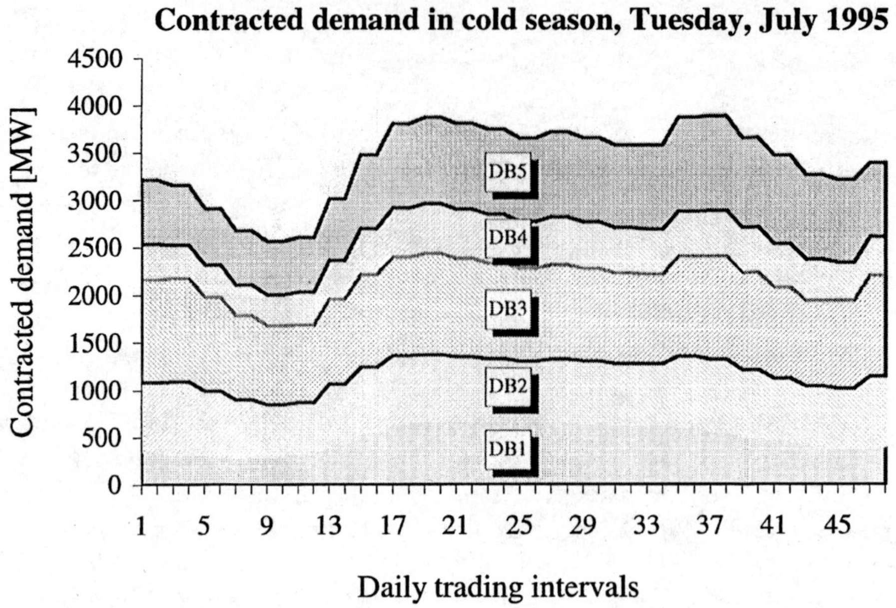


Fig.2 Vesting contracts for a working day in a cold season in July 1995

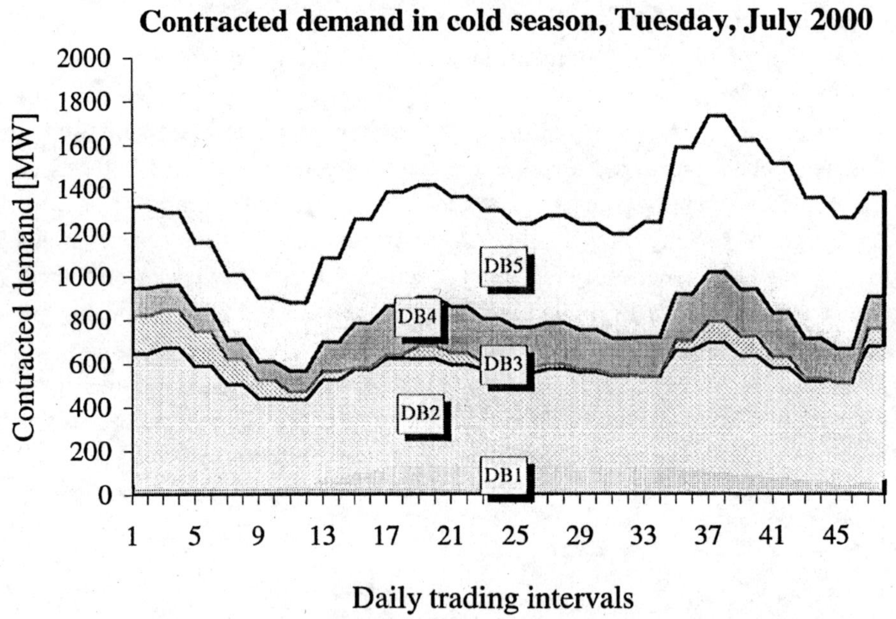


Fig.3 Vesting contracts for a working day in a cold season in July 2000

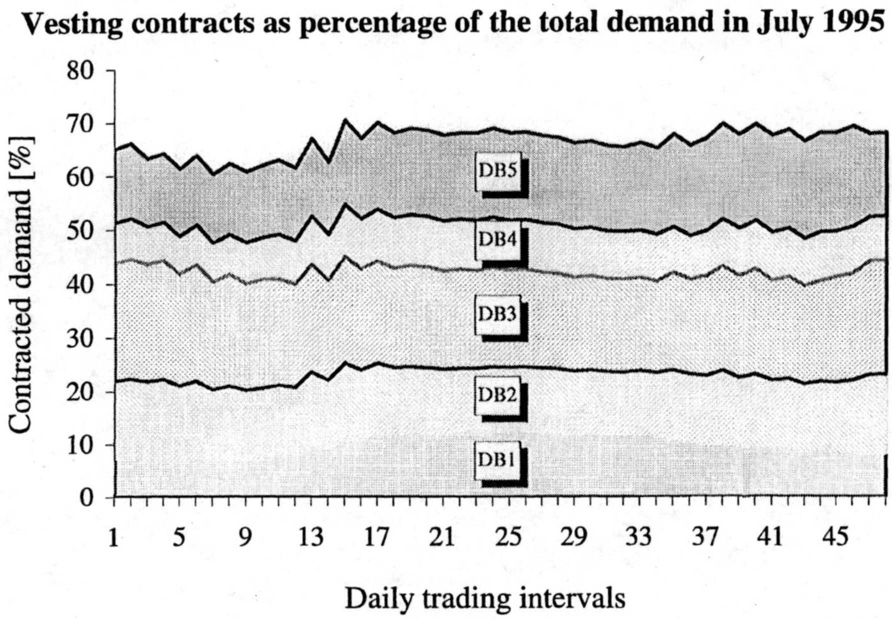


Fig.4 Vesting contacts as percentage of the total demand in July 1995

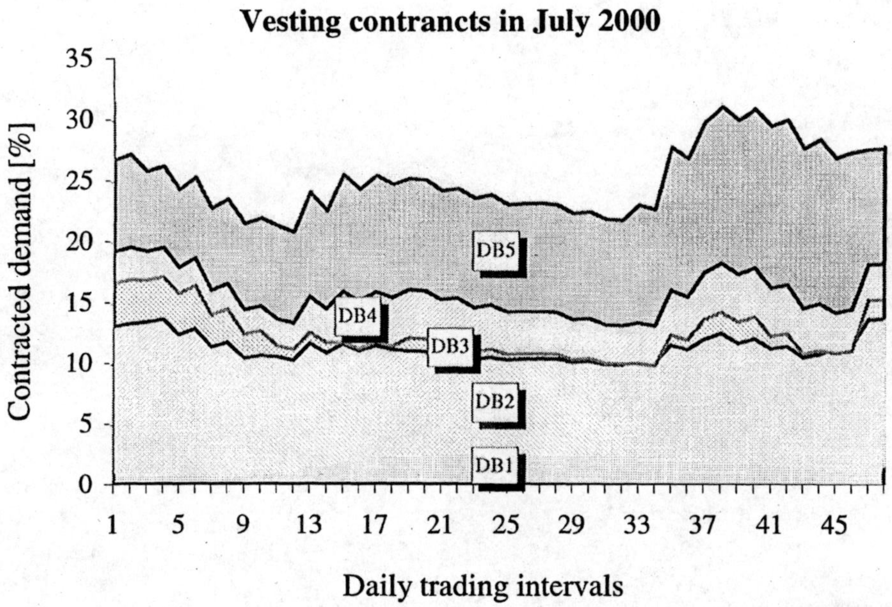


Fig.5 Vesting contacts as percentage of the total demand in July 2000

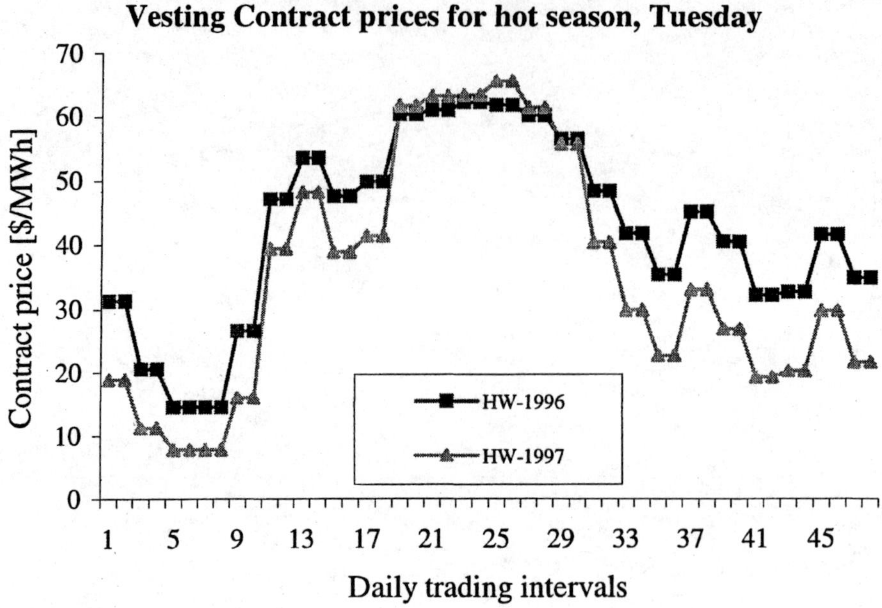


Fig.6 Vesting contract prices in hot season for a working day

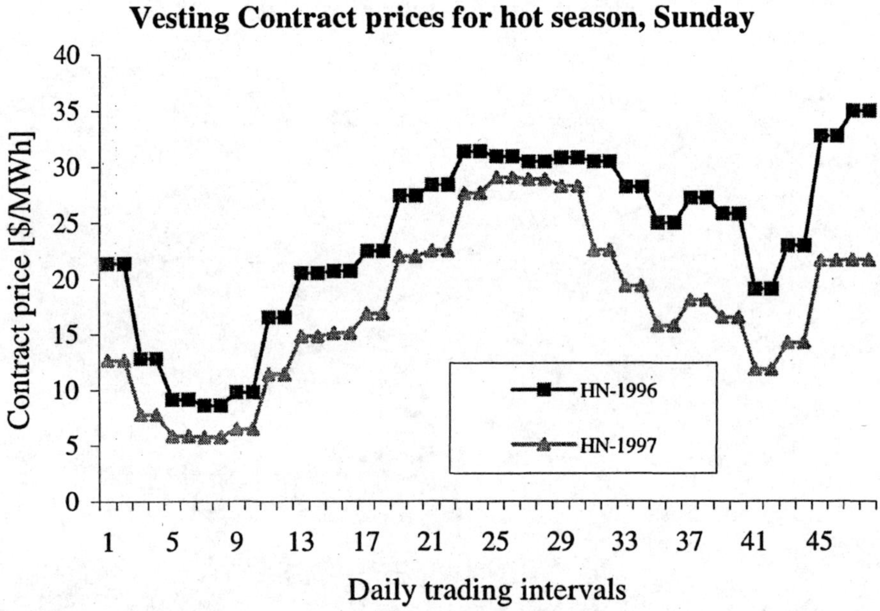


Fig.7 Vesting contract prices in hot season for a non working day

## 3 BILATERAL CONTRACTS

Bilateral Hedge contracts, long and short term, are voluntary contracts arranged between energy retailers and power stations. Market participant may reveal amount of electricity contracted by bilateral contracts; however the price in such contract is always confidential. Some estimates indicate the electrical energy prices in bilateral contract vary between $20 and $30 per MWh. An example of contract prices published by Electricity Week, Issue – December 9, 1997 are shown in Figure 8. Middle term contract prices rise in December 1997 in expectation of shortfall of supply in hot summer months (December – March).

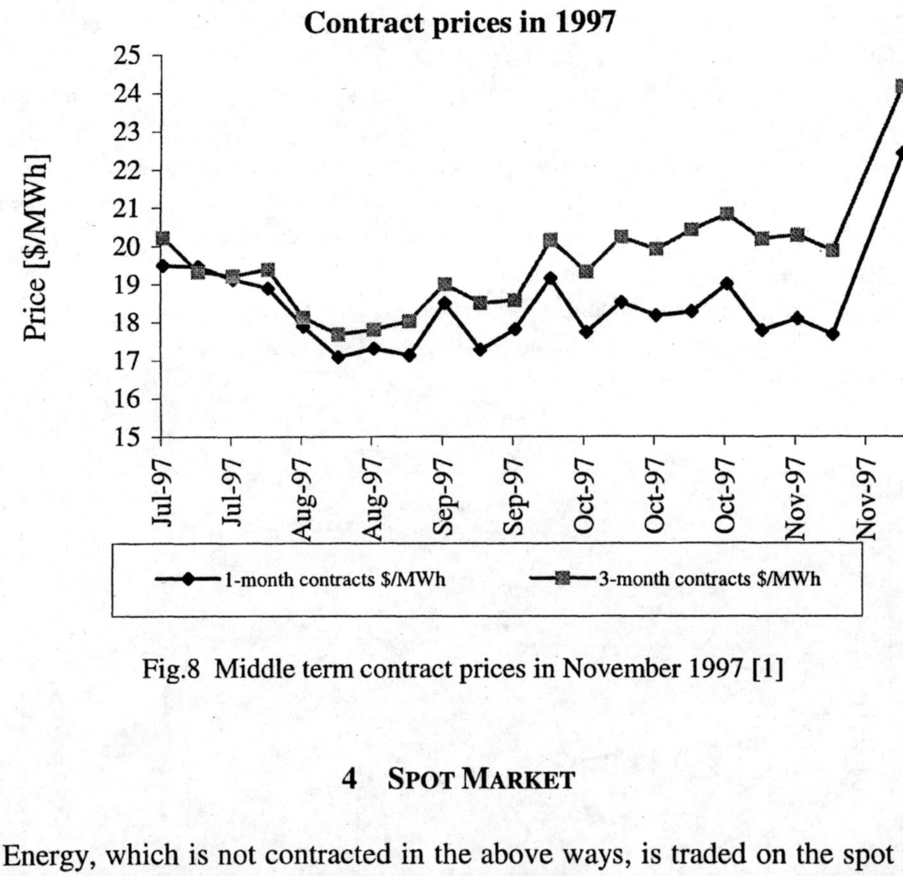


Fig.8 Middle term contract prices in November 1997 [1]

## 4 SPOT MARKET

Energy, which is not contracted in the above ways, is traded on the spot price market. Spot prices result from the balance between supply and demand in a bidding system. Currently, participants provide their bids in 10 price bands in 0.5-hour trading intervals. The market manager calculates a spot price in 5-minute periods. Vesting and hedge contract payments are calculated in 0.5-hour intervals by averaging 5-minute spot prices.

The analysis of spot prices and energy demand covers two periods:

- from July 1994 to May 1997 (the VicPools)
- from May 1997 to January 1998 (the NEM1).

## 4.1 DEMAND AND SPOT PRICES IN THE VICPOOLS AND THE NEC

The analysis of spot prices over 3.5 years of the VicPool operation shows that average spot prices have decreased significantly despite some price oscillations and slight increase in energy demand (about 3%), as seen in Fig. 9. In the beginning of the VicPool (July 1994), the average spot price was about $40/MWh, while at the end of April 1997 the average energy price was almost 50% lower. Although demand for electricity varies with time of the day, days of the week and seasons, the average pool demand remains practicable stable and for Victoria accounts for about 4500 MW. On the other hand, oscillations of spot prices, especially during the first two years of market operations, were significant and not necessary justified by the existing market conditions. This may be explained by the lack of experience of market participants in open electricity market operation. Generally, two main factors contributed to the reduction in spot prices. First, the existence of vesting and bilateral hedge contracts in which energy producers were able to recover their fixed costs. These contracts allow Generators to sell energy excess charging only variable costs. Secondly, improved, more efficient organisation introduced by new power station owners contributed to the decrease in spot prices.

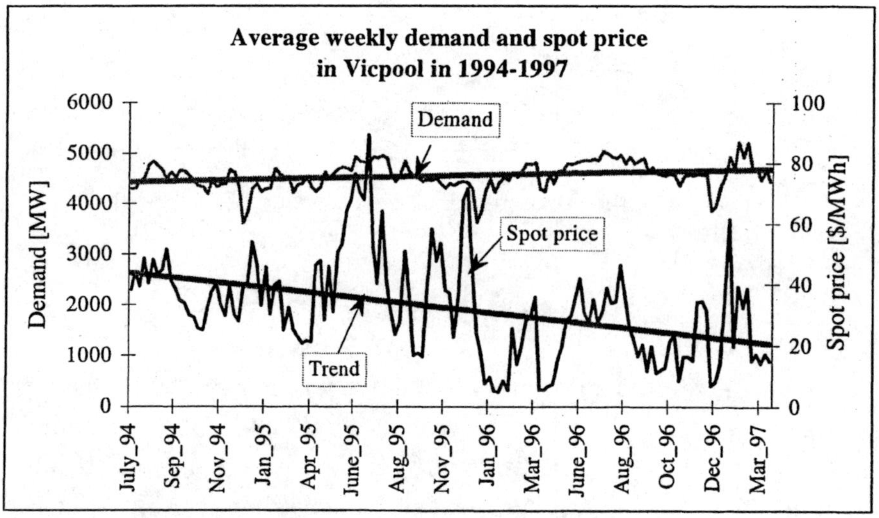


Fig. 9 Average weekly demand and spot prices in VicPool between July 1994 and April 1997

After introduction of the National Electricity Market (NEM) in May 1997, spot price oscillations has been reduced, preserving the falling tendency of spot prices, as seen in Fig. 10. However, some unusual events can lead to significant market instability. For example, increased demand caused by high temperatures along with transmission limitations at the end of November 1997 resulted in the dramatic temporary increase in spot prices.

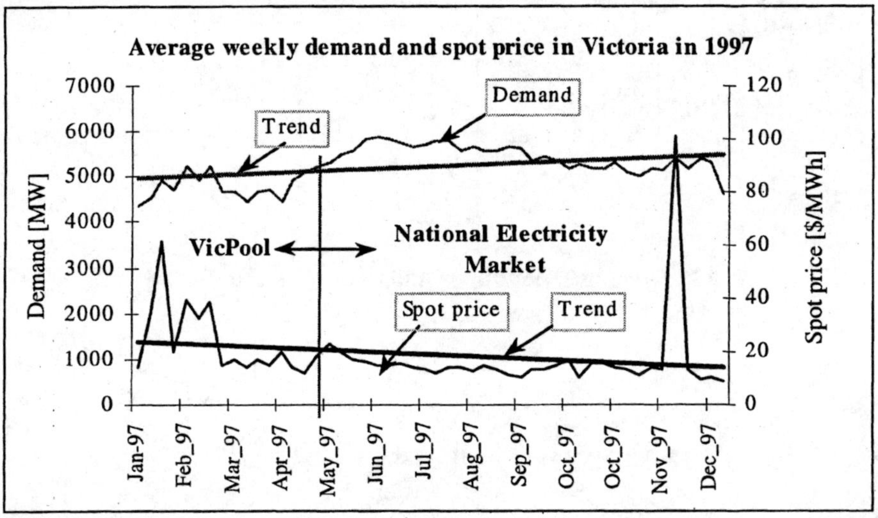


**Fig. 10 Energy demand and spot prices in the VicPool and the NEM in 1997**

## 4.2 Average Weekly and Monthly Demand and Spot Prices

The analysis of average weekly and monthly spot prices for Region 1 (Victoria and South Australia) and Region 2 (New South Wales and Australian Capital Territory) reveals similar levels of spot prices and their standard deviations, as seen in Fig. 11 and Fig 12. It can be observed that the spot prices for Region 1 are slightly lower than for Region 2. This is a result of low costs of energy generation in Victoria, where bulk of electricity is supplied from brown coal fuelled power stations.

In normal conditions, a link between two regions allows to transfer sufficient amount of energy to reduce spot price deviations. However, in the case of large demand and/or severe network limitations, capacity of the link does not allow the adequate energy transfer; thus, the level of spot prices in the both regions can differ significantly.

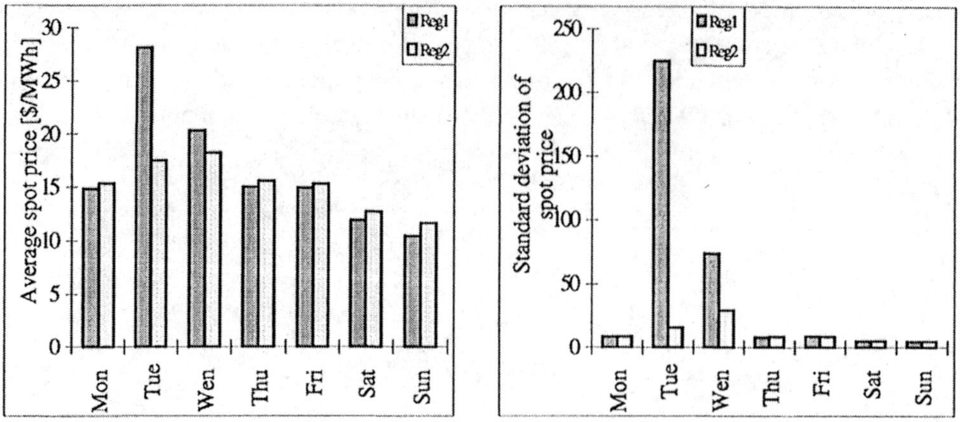


Fig. 11 Weekly average spot prices and their standard deviation in 1997

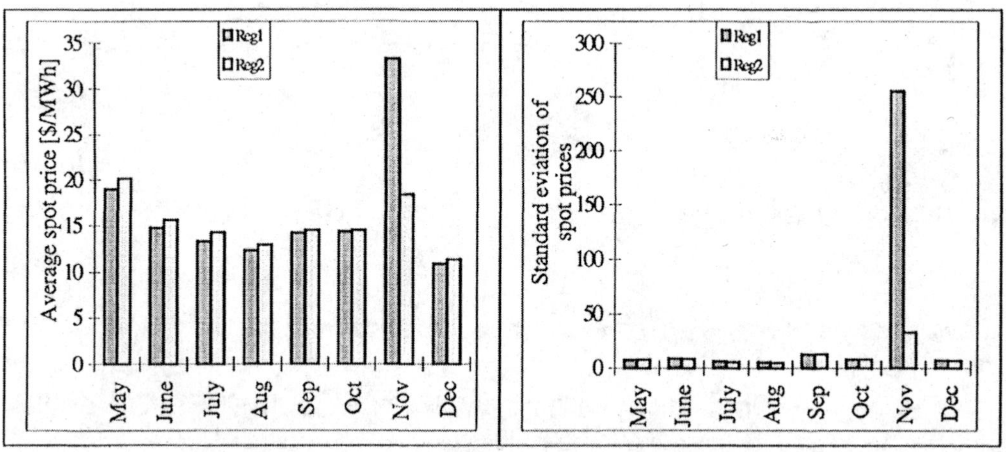


Fig. 12 Monthly average spot prices and its standard deviation in 1997

Average pool demand in both regions is relatively stable, and in 1997 accounted for about 5000 MW in Region 1 and 7.500 MW in Region 2, as can be seen in Figures 13 and 14. Standard deviations of demand are below 10% in Region 1 and 13% in Region 2.

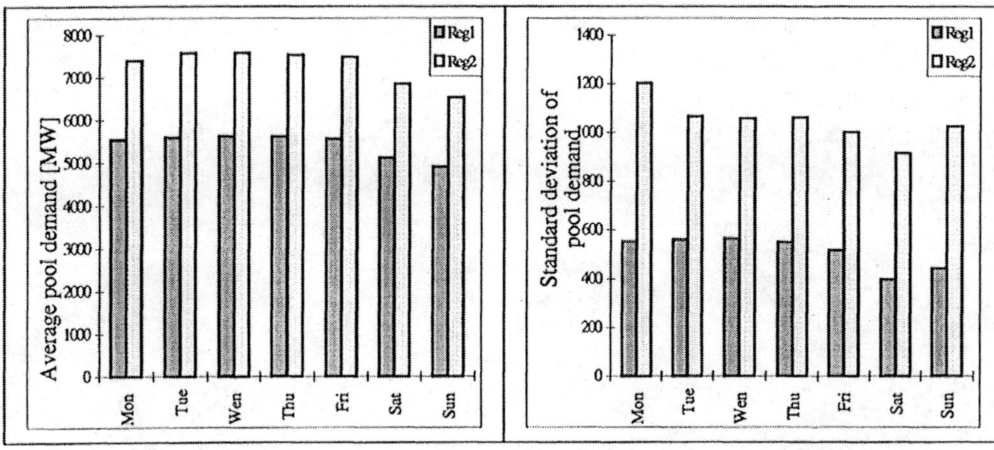


Fig. 13 Weekly average pool demand and its standard deviation in 1997

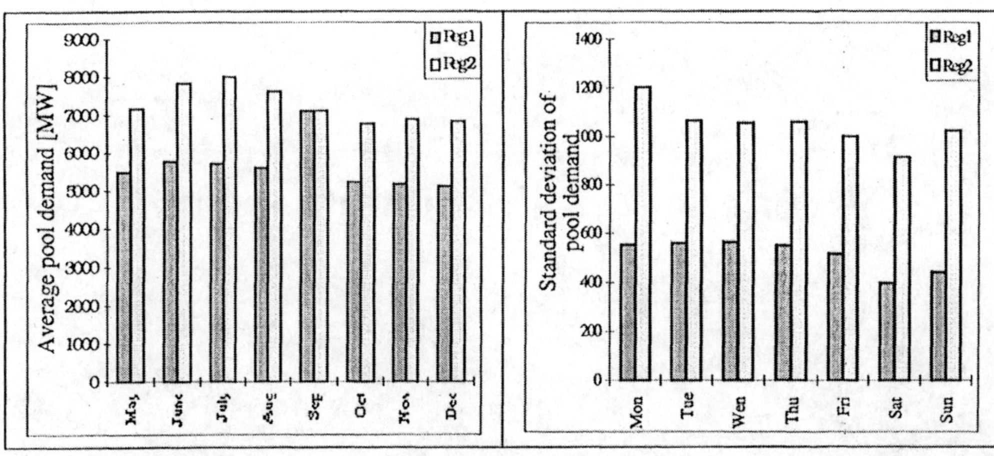


Fig. 14 Monthly average pool demand and its standard deviation in 1997

### 4.3 WHEN SPOT PRICES REACH VOLL

Significant abnormality in spot prices occurred for Region 1 on Tuesdays (Fig. 11) and November (Fig.12) relates to $25^{th}$ of November 1997, when Melbourne had the hottest November day for 86 years with the temperature reaching 40.3°C. The high demand for electricity along with a scheduled transmission outage in the Snowy Hydro - Victoria and two offline 500 MW units brought the Victorian market to the official ceiling, the Value of Lost Load (VoLL) that accounts for $5,000, in one five-minute interval [1].

As a consequence, the average 0.5-hour interval price jumped from \$8.63/MW to \$4,814/MW, as seen in Fig. 15. Although the transmission was reconnected on Wednesday, the panic among the Market Participants resulted in very high spot prices also for that day with the maximum value reaching \$2,450/MW (see Fig. 16). On Thursday, spot prices returned to their normal low values. Electricity demand and spot prices during the "black week" in November 1997 are shown in Fig. 17.

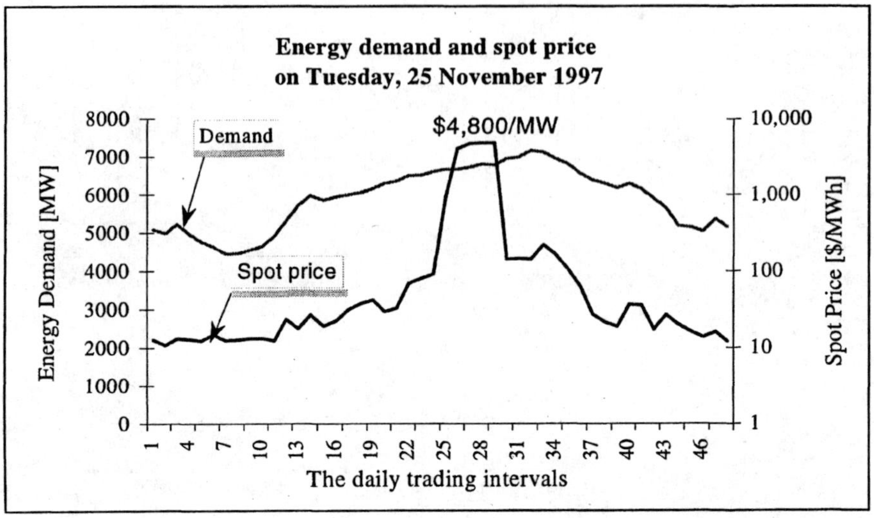


Fig. 15 Demand and spot price on Tuesday, 25th of November 1997

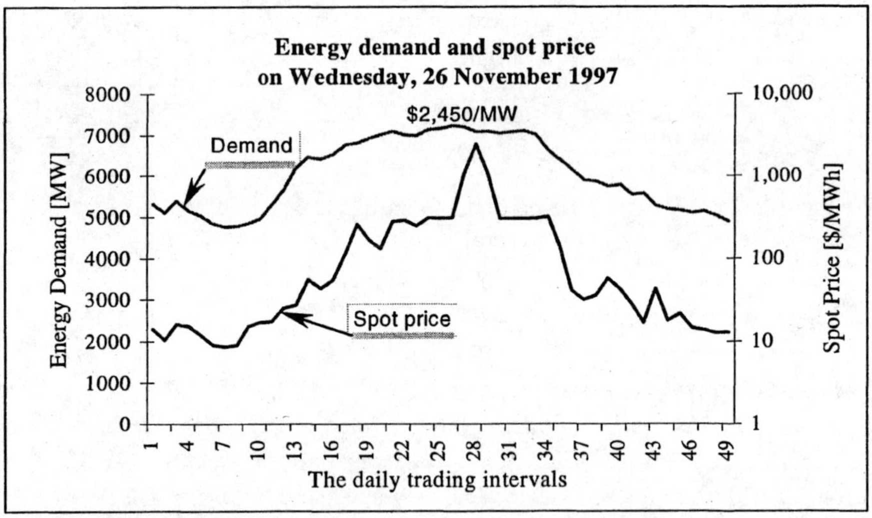


Fig. 16 Demand and spot price on Wednesday, 26th of November 1997

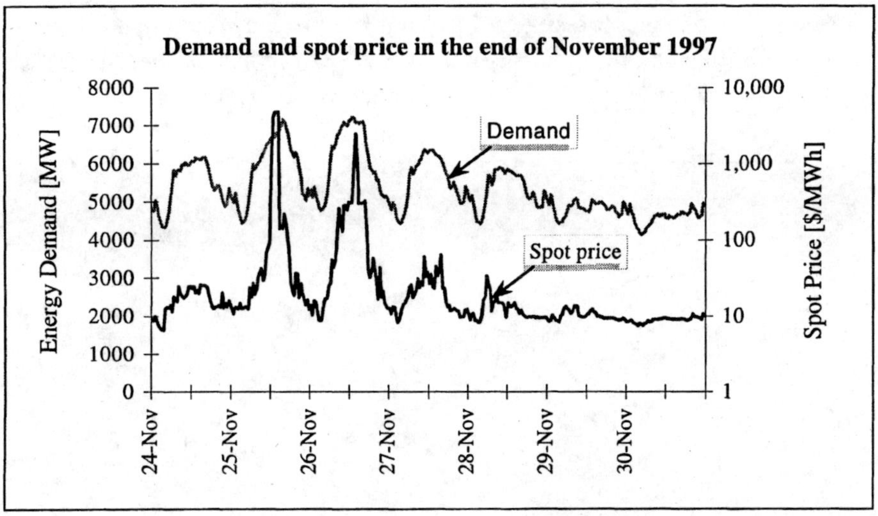


Fig. 17 Demand and spot price for Victoria in the last week of November 1997

During the "black week", some power stations noted huge income, especially Generators with low coverage by vesting and hedge contracts. For example, on Tuesday, 25$^{th}$, due to Electricity Week, Southern Hydro earned $4.3 millions compering to usual trade per day equal to $0.15 million [1]. Vesting and hedge contracts partly reduce the expose of retailers to market volatility. However, the experience from these two untypical days justifies a need for improved protecting financial instruments. It is characteristic that the number of trades on the futures exchange in Sydney went from 0 to 67. It is common perception that the experience from the 'black week' will result in larger number of hedge contracts and higher contract prices.

## 5. Correlating Spot Prices

Spot prices are influenced by several factors including electricity demand, system capacity, generation surplus, and energy transfer. The correlation between spot prices and the above-mentioned factors was investigated for both the VicPool and the National Electricity Market (NEM). The results are presented for May 1997, the first month of the National Electricity Market operation.

### 5.1 Spot Prices and Electricity Demand

Spot prices depend in great degree on electricity demand; therefore, it is necessary to investigate functional relations between these two variables. Correlation of average spot prices and electricity demand for Region 1 in May 1997 is shown in Fig. 18. The analysis carried out for VicPool and for other months of the NEM indicates similar relationships. It can be observed that the spot price patterns fit the electricity demand quite accurately.

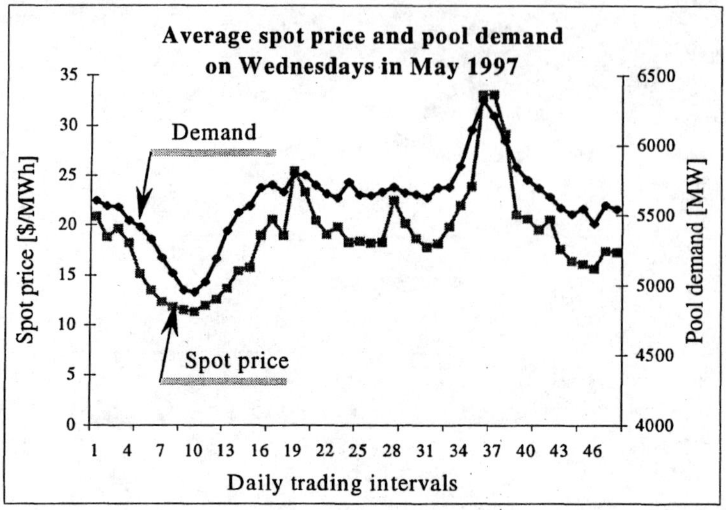


Fig. 18 Average spot prices and electricity demand for Region 1 on Wednesdays in May 1997

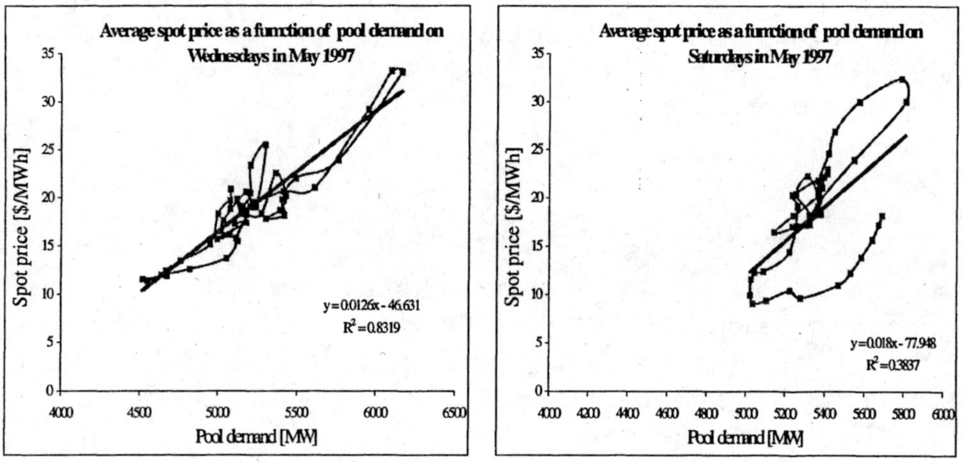


Fig. 19 Average spot price as a function of pool demand for Region 1 on Wednesdays and Saturdays in May 1997

Spot prices can be directly expressed as a function of pool demand. This allows drawing a trend line and investigating the accuracy of trend approximations. The trend relations between spot prices and pool demand for Wednesdays and Saturdays in May 1997 are shown in Fig. 19. The data used is the same as in Fig. 18 for the time dependent function. It can be noticed that the linear trend can quite accurately approximate a functional relationship between spot prices and pool demand for work days; however, spot prices during non-work days show significant deviations from the trend lines.

## 5.2 Spot Prices and Available Capacity

Second parameter that can have the influence on spot prices is the available capacity. Generators when providing offers to the pool manager have to define not only spot price bands and the amount of power offered within these bands but also they provide the information on units' availability. The relations between available capacity and spot prices for Region 1 on Wednesdays in May 1997 are shown in Fig. 20.

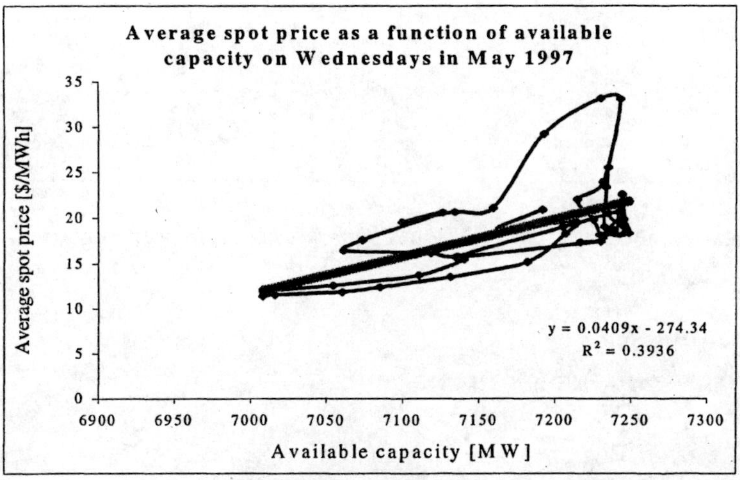


Fig. 20 Average spot prices and available generation capacity on Wednesdays in May 1997

It can be observed that when the more generation capacity is available, the hither is the price, because units that are more expensive (gas turbine generating units) are brought into operation. However, the available capacity does not provide the direct information on spot price patterns, since these patters are more affected by the pool demand.

## 5.3 Spot Price and Generation Surplus

Generation surplus is calculated as a difference between available generation capacity and the demand forecasted. The surplus of generation reduces spot prices because generators tend to sell electrical energy on lower prices when facing stronger competition. The analysis of a functional relationship between spot prices and generation surplus shows the trend with a negative slope. During workdays, a linear trend provides more accurate approximation of a relation between spot prices and generation surplus compering to approximation for non working days. An example for Wednesdays and Saturdays derived from the data for May 1997 is shown in Fig. 21.

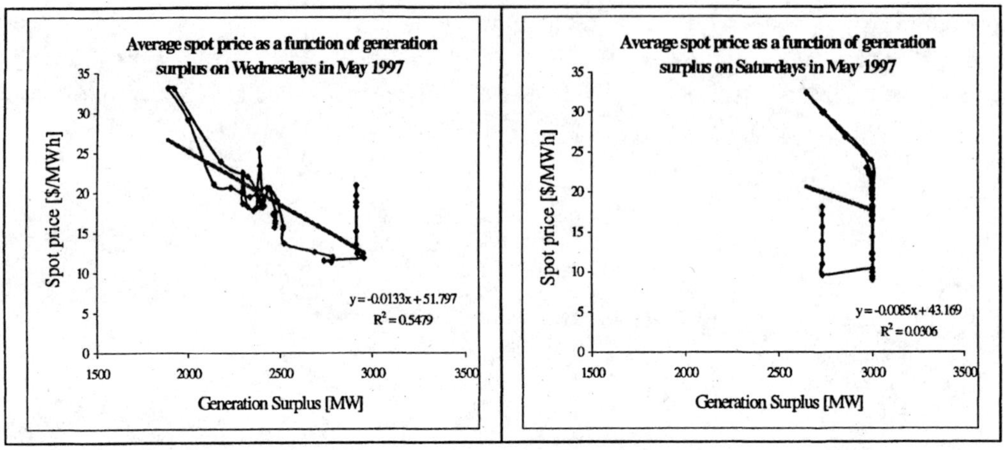


Fig. 21 Relation between spot prices and generation surplus for Region 1 in May 1997

## 5.4 Spot Prices and Energy Transfer

A link between Region 1 and Region 2 allows the transfer of energy to reduce differences in spot prices between the regions. Although the capacity of this link is limited, it plays an important role in reducing the overall cost of electricity. The analysis of energy transfer and spot prices has been carried out since May 1997, that is, from commencement of the National Electricity Market. However, the relationship derived do not provide straight functional relations, revealing functions that are more complicated, see Fig. 22.

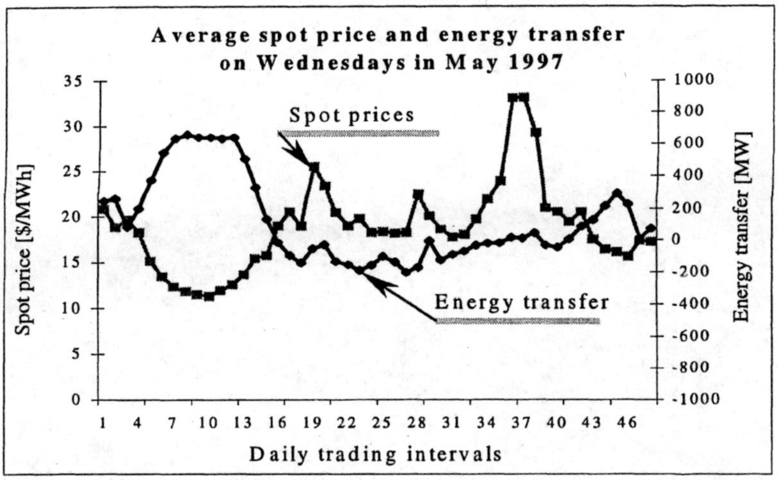


Fig. 22 Average spot price and energy transfer between Regions1 and 2 for Wednesdays in May 1997

The relations between spot price and energy transfer should be analysed separately for three characteristic periods:

1. ***Night valley***

During low demand in both regions, the thermal power stations in Region 2, which are using black coal, reduce generation to minimum due to high variable cost. Thermal power stations in Region 1, which use brown coal as a fuel, have low variable cost. These power stations, located in La Trobe Valley, have enough capacity to cover electricity demand in Region 1 and export electrical energy to Region 2. This can explain why spot prices in Region 1 are low with large export to Region 2.

2. ***Morning and afternoon demand***

In this period, spot prices and export varies, showing small positive correlation. The increase of export to Region 2 results in small increase of spot prices in Region 1.

3. ***Evening demand peak***

In this period, prices in both regions are increasing as the result of dispatching more expensive generation units. Energy export oscillates around zero value when both regions have similar spot prices.

A direct application of functional relationships between energy transfer and spot prices does not provide important information for spot price forecasting, as seen in Fig. 23. This is a result of complex relationships with varying dependence during three characteristic periods. Trend approximations should be carried out separately for each period defined above.

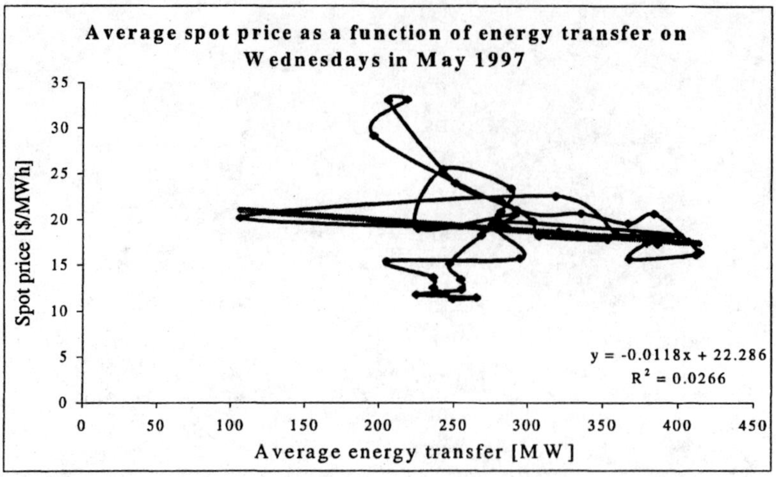


Fig. 23 Trend approximation between spot prices and energy transfer for Wednesdays in May 1997

As can be seen from Fig. 24, during off-peak hours, spot prices in Region 1 are lower when energy transfer to Region 2 increases. During working hours, when the energy is imported from Region 2, the influence of energy transfer on spot prices is less significant.

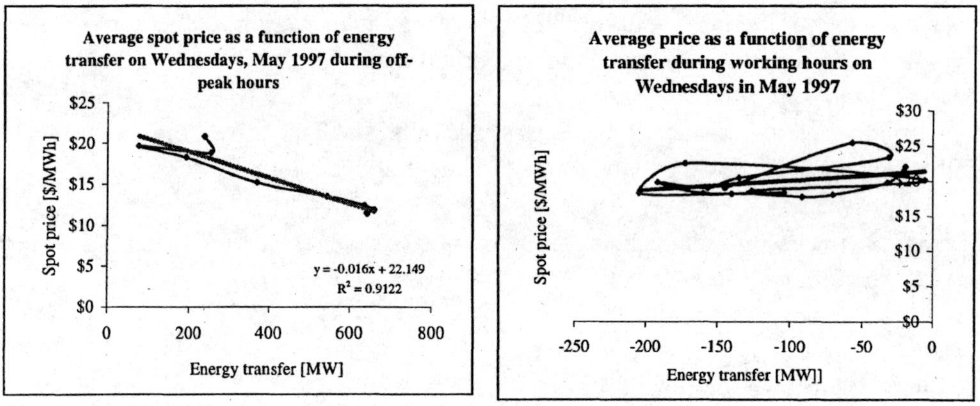


Fig. 24 Energy transfer and spot price trends for off peak and working hours periods.

## 6 Conclusions

Demand for electricity in the Vicpool and the National Electricity Market is relatively stable, showing the increasing trend with the slope of 3-5% depending on distribution business. It is expected that electricity consumption may decrease when the Australian economy will be effected by the crisis in Asia. However, electricity demand is expected in the worse case to stabilise on the current level.

Spot prices for electrical energy are very low with average values oscillating around \$10 to \$12/MWh. This is caused by high prices in vesting and bilateral contracts. These contracts allow power stations to recover fixed costs of generation. The excess of energy is trading on the spot market on low prices to recover variable costs. In year 2001, when all customers become contestable customers, vesting contract will expire. This may force power station to increase the prices offered in the spot market. Moreover, some power stations are enough large that they may drive spot prices up as the demand cannot be covered with their significant commitment.

Although spot prices are varying, there are clear relationships of prices to several *spot market parameters. This allows to develop a methodology of electricity contracting* to reduce risk of market exposure. Monash University is carrying out a research project leading to the Decision Support System which will assist in electricity contracting. Influence of market uncertainty on market risk is reduced by the application of fuzzy sets techniques.

## Reference

1 Electricity Week, Tuesday News Analysis Edition: December 9, 1997, Vol. 21, No 2

****** 15 ******

# TOWARDS THE NATIONAL ELECTRICITY MARKET

**WLADYSLAW MIELCZARSKI AND GRAZYNA MICHALIK**
Monash University, Department of Electrical and Computer Systems Engineering
Melbourne, Australia, Email: Wlad@basil.eng.monash.edu.au

**Summary:** The first stage of the National Electricity Market (NEM) known as NEM1 was introduced in May 1997. Currently the market consists of two main regions comprising three states: New South Wales, Victoria and South Australia, and the Australian Capital Territory. The paper presents the structure and operation of the National Market, dispatch and security issues, settlements, inter and intra regional loss calculation and the regulation of the competitive market.

## INTRODUCTION

The National Electricity Market, which was introduced in May 1997, is regulated by the National Electricity Code. The code has its genesis in the Special State Premiers' Conference in October 1990 and July 1991, which led to the formation of the National Grid Management Council. The council published in October 1993 the discussion paper in which it recommended a range of regulatory arrangements for the national electricity grid. It was indicated that the operation of the national grid would be regulated by way of Code of Conduct [1].

The development of the national grid and the Code is overseen by a national regulator such as the Australian Competition and Consumers Commission and the states regulatory bodies like the Office of the Regulator-General in Victoria to be consistent with the state regulations. The States of New South Wales (NSW), Victoria, Queensland, South Australia and the Australian Capital Territory have agreed to the enhancement of cooperative legislation for the implementation of the regulatory arrangement. This legislation is called the National Electricity Law. Two institutions have been established:

the National Electricity Code Administrator (NECA) and the National Electricity Market Management Company to develop and introduce the NEC to operation of the National Electricity Market. NECA is the corporate body responsible for administrating the Code. NEMMCO is responsible for operating and administering the market in accordance with the Code.

The main streams of the reform process are shown in Fig.1. There are two streams of activities:

- Regulation
- Implementation

The regulation leading to the National Electricity Market (NEM) creation and operation started from the National Electricity Law that founded the basis for restructure of the electricity supply industry.

The National Electricity Code (NEC) developed by the National Electricity Code Administrator (NECA) provides rules for operation of the National Electricity Market. After the introduction of the NEM, NECA will be overlooking the operation of the NEM and the compliance with the code rules. NECA will also judge disputes between market participants. The implementation of the reform and operation of the National Electricity Market is supervised by NEMMCO [2].

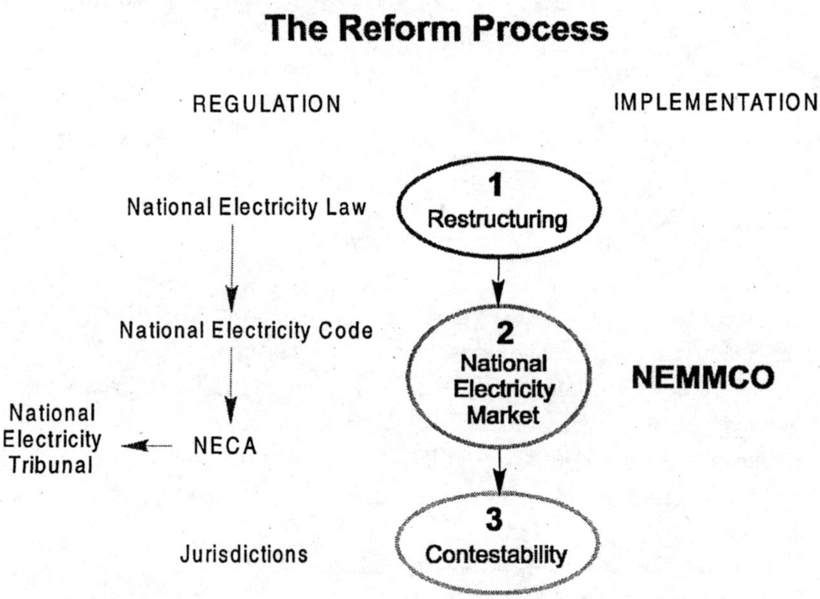


Fig. 1 Reform of the Australian electricity supply industry

## 2 FRANCHISE AND CONTESTABLE CUSTOMERS

The creation of the NEM leads to contestability, that is, customers have rights to select their electricity retailer and negotiate financial conditions of supply. Table 1 presents the customer contestability timetable in Australian States [1]. As seen from Table 1, the most advanced reforms has been undertaken in Victoria, where the first 47 large customers became contestable in December 1994. In January 2001, all customers in Victoria, including residential customers, will be legible to select their electricity retailers.

Table 1 Customer contestability timetable by participating jurisdiction

| Customer Size | Victoria | South Australia | New South Wales | Australian Capital Territory | Queensland |
|---|---|---|---|---|---|
| Above 5 MW or 40 GWh per annum | December 1994 47 sites | April 1998 25 sites | October 1996 47 sites | October 1997 5 sites | January 1998 43 sites |
| Above 1 MW or 4 GWh per annum | July 1995 330 sites | July 1998 125 sites | April 1997 660 sites | March 1998 40 sites | January 1999 346 sites |
| Above 750 MWh per annum | July 1996 1,500 sites | January 1999 600 sites | July 1997 3,500 sites | May 1998 274 sites | Not classified |
| Above 160 MWh per annum | July 1998 5,000 sites | January 2000 2,400 sites | July 1998 10,800 sites | July 1998 781 sites | January 2000 6,317 sites |
| All customers | January 2001 1,957,300 sites | 1999* 700,000 sites | July 1999* 2,7000,000 sites | July 1999 125,000 sites | January 2001 1,407,000 sites |

* To be reviewed

Graphically, a choice of a retailer is shown in Fig. 2. Before the introduction of competitive markets, all customers were Franchise Customers, that is, they paid for electrical energy due to the tariffs prescribed by the supply authority. The reforms to electricity market caused that:

- Electricity tariffs are set up by suppliers, but they have to be approved by the Regulator in order to retain energy prices below presumed level.
- A Contestability timetable was set allowing the first large customers to select their retailers or join the wholesale market. Gradually, all customers will become contestable customers.

A contestable customer can join the wholesale market by registration with NEMMCO as Market Customers. Only very large customers take this option, since the cost of participation in the electricity pool is significant. Retailers usually do not own the distribution network [1].

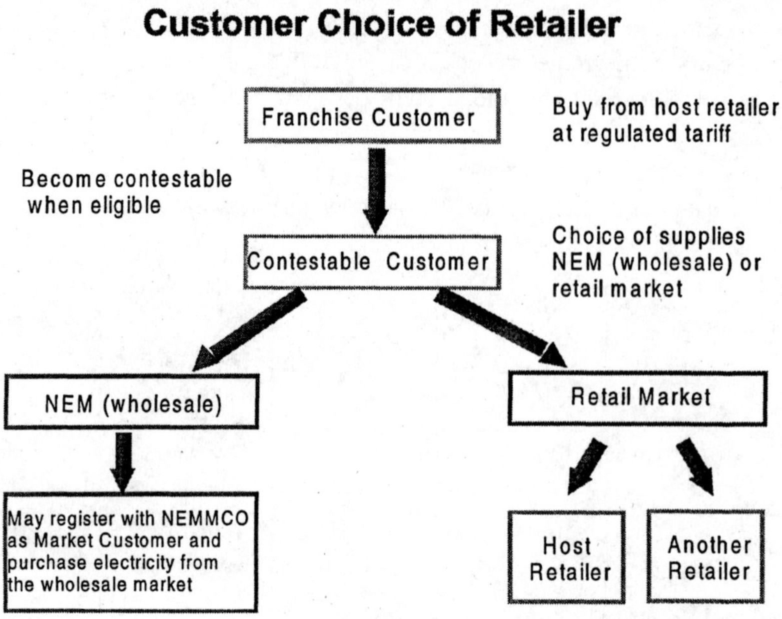


Fig. 2 Customer choice of supplier

## 3 Structure of the National Electricity Market

The National Electricity Market is a wholesale market for the supply and purchase of electrical energy with an open access to transmission network. Currently, in the first stage, which started in May 1997 as NEM1, the market comprises two main regions

- Region 1 – Victoria and South Australia (SA)
- Region 2 – New South Wales (NSW) and Australian Capital Territory (ACT)

The final structure of the market is supposed to comprise five regions:

- South Australia
- Victoria
- Snowy Mountains Generation Scheme
- New South Wales with ACT
- Queensland

NEMMCO will manage and facilitate the wholesale electricity market, while NECA will supervise, administer, and enforce the National Electricity Code (NEC).

Two basic components of the electricity pool are centrally coordinated:

- Generator dispatch process
- Spot price market

Generators compete by providing bids to NEMMCO that determine price bands, unit availability, and dispatch offers. Market customers may submit dispatch bids, comprising prices and associated quantity of demand they wish to be scheduled in the dispatch process. Generator bids may contain a self dispatch level. In this case the price for self dispatch power must be negative, as seen in Fig. 3 [1].

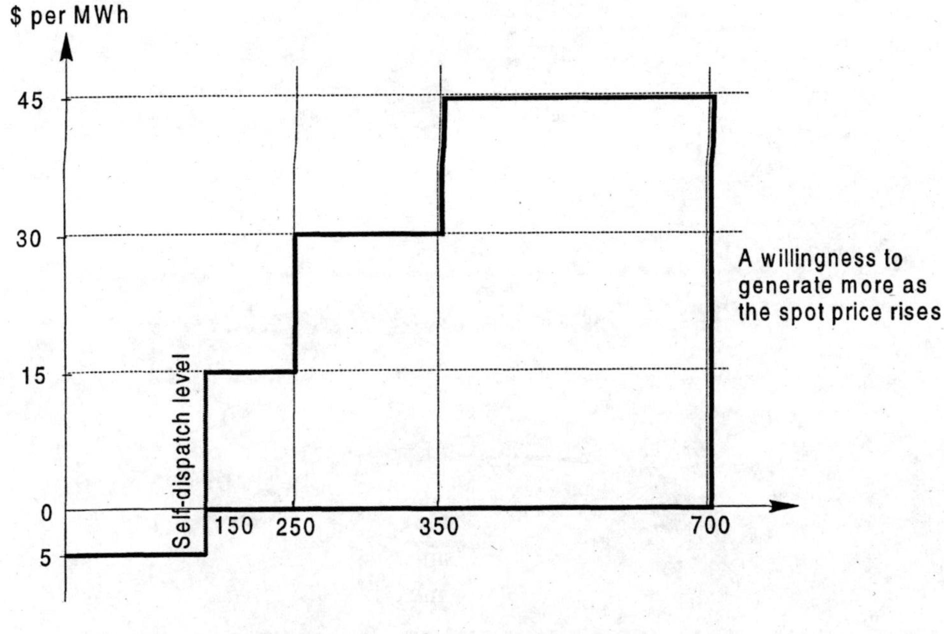


Fig. 3 Typical Generator offer

There are four main categories of Generators [1]:

- Scheduled generator – a Generator with the nameplate rating above 30 MW at one site. It must be scheduled by NEMMCO.
- Non-scheduled Generator – a Generator with a nameplate rating of less than 30 MW at one site. It is not required to be scheduled by NEMMCO.
- Market Generator – a Generator who sends out electrical energy which is not entirely purchased by the local retailer. This Generator is required to send its output through the spot market.

- Non-market Generator – a Generator whose sent out electrical energy is purchased entirely by a local retailer or customers located at the same network connections.

Additionally to Generators, two other Market Participants must be registered with NEMMCO (see Fig. 4):

- System Operator
- Distribution System Operator

A System Operator is a person who is appointed by NEMMCO as its agent under the Code to manage operation of the power system. Distribution System Operator is a person who is responsible for managing the operation of a distribution network and directing its operation during emergency. The Distribution System Operator manages the transfer of electricity throughout the distribution network to the end customers [1] [4], see Fig. 4

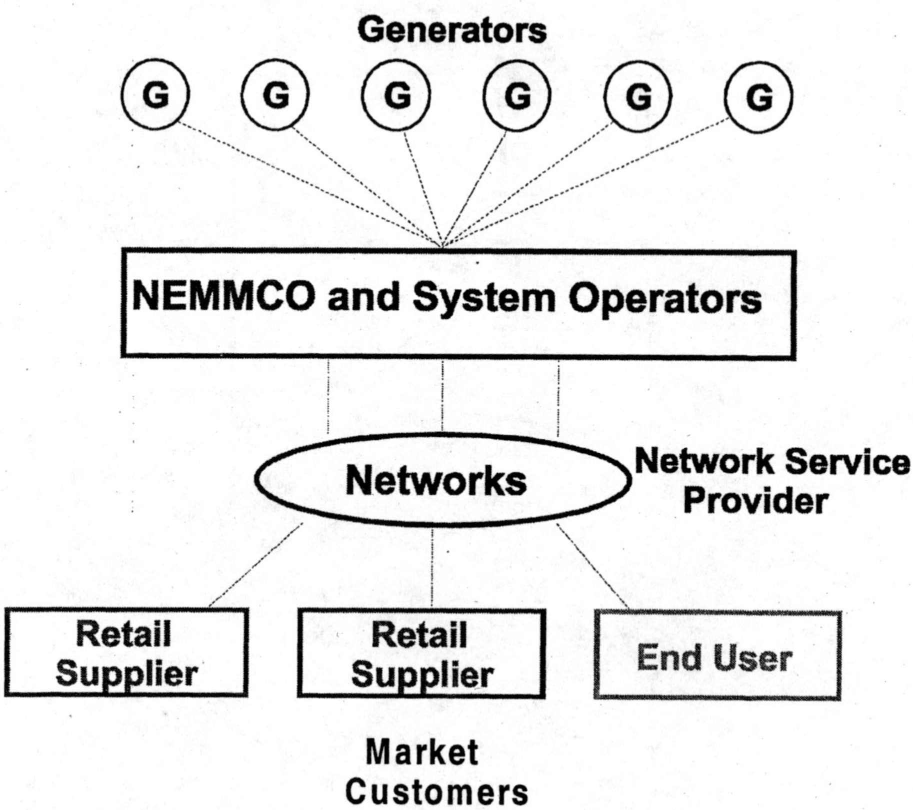


Fig. 4 Structure of the National Electricity Market

Transfer of electrical energy via the pool is shown in Fig. 5. Generators provide electricity to pool from which Retailers and large energy users purchase energy on the spot market. Small energy users purchase electricity from electricity Retailers. Participants of the wholesale electricity market can enter into various forms of hedge contracts in order to reduce risk exposure to very high (Retailers) or very low (Generators) spot prices. In some cases, third parties, including banks or other financial institutions may be involved in hedge contracts.

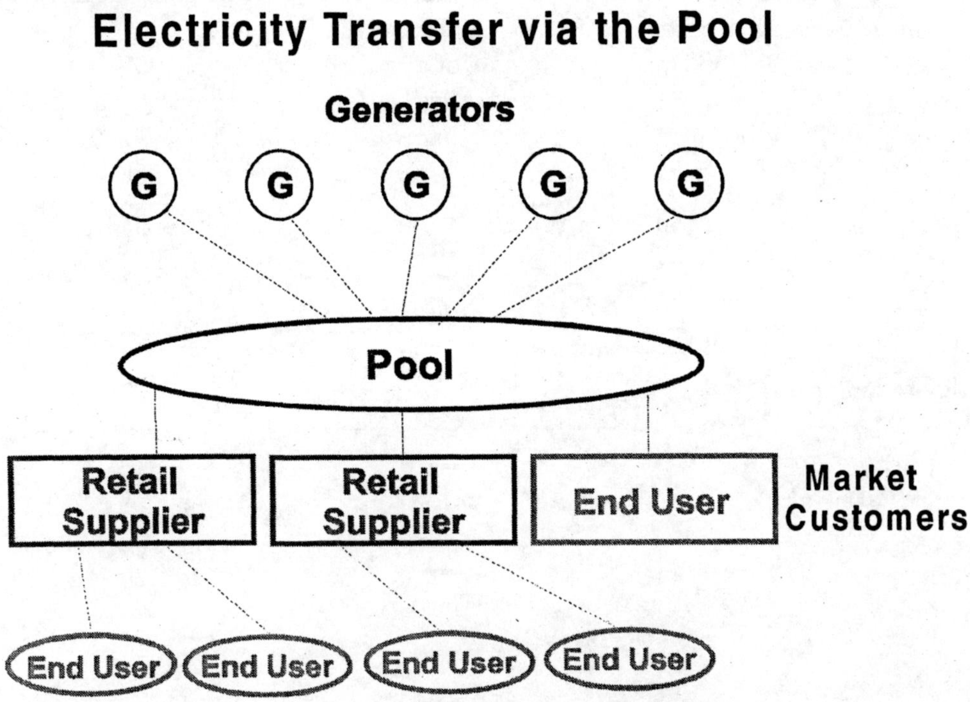


Fig. 5 Electricity transfer via the pool

## DISPATCH AND SECURITY CENTRES

NEMMCO will operate two National Dispatch and Security Centres to manage pool operation. NEMMCO is also responsible for procurement and management of the computer systems and data communication infrastructure. The computer system, called the NEM System, consists of the Market Management and Scheduling, Pricing and Dispatch Systems [1] [4], as seen in Fig. 6.

***Market Management Systems*** comprise:

- bidding and reporting systems, which accept bids and provide market information

- settlement system, which prepares invoices for Generators and Retailers based on metered electrical energy.

Scheduling, Price and Dispatch Systems (SPD) comprise:

- dispatch system, which uses bids to schedule Generators
- spot price calculation system

Spot prices are determined in 5-minute periods and next they are averaging to 0.5-hour periods. The payment for Generators is based on 5-minute spot prices, while vesting and hedge contracts are formulated for 0.5-hour trading periods. This requires the averaging of 5-minute spot prices.

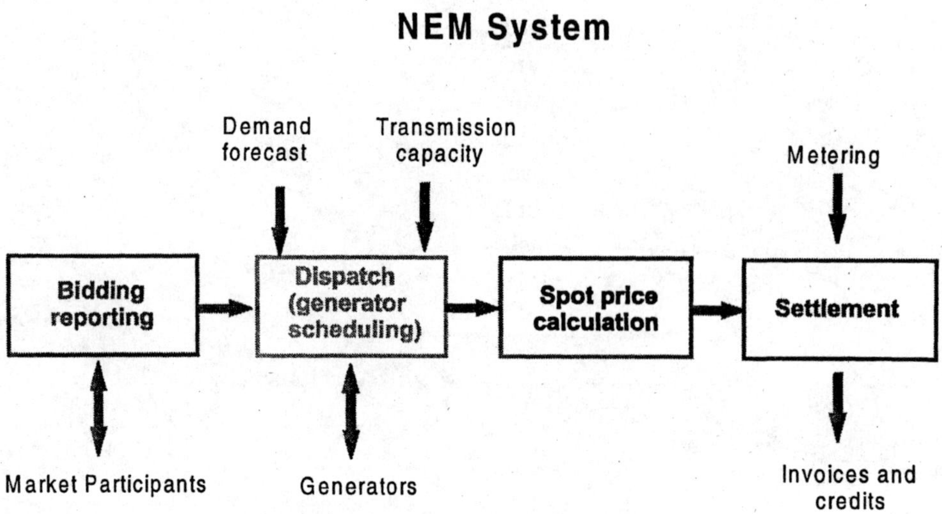


Fig.6 Structure of NEM system

## 4- SETTLEMENTS

The main structure of the settlement process is shown in Fig. 7. NEMMCO will issue a weekly spot market account to each Generator and Retailer [1] [3].

A Generator spot market account comprises:

- payment for generated electrical energy
- payment for ancillary services if provides
- pool fees paid to NEMMCO and NECA

A Retailer spot market account comprises the following items:

- charges for electrical energy consumption
- charges for ancillary services
- pool fees owed to NEMMCO/NECA

## 6 NATIONAL MARKET REGIONS

Operation of the National Electricity Market is limited by the capacity of links between States and generating schemes. A link between Victoria and South Australia (SA) allows the transfer of 500 MW towards SA and 250 MW in the opposite direction. This accounts for less than 10% of the Victorian pool demand.

A link between New South Wales (NSW) and Victoria includes the energy transfer from Snowy Mountain Generation Scheme (Snowy). Several hydro power stations were constructed in Snowy Mountains in the forties and fifties. This common investment of NSW and Victoria entitles Victoria to 1100 MW from Snowy, while the NSW share accounts for 2700 MW. However, the existing capacity of the link between NSW and Victoria limits energy transfer between these two regions. During peak hours, the 1500 MW capacity is used partly by the Victorian entitlements from Snowy (1100 MW), leaving only about 400 MW to transfer energy to Victoria. Similarly, the transfer capacity to NSW is limited. Currently, Victoria exports some amount of energy to NSW during off-peak hours. The limited capacity of links is especially visible during periods of high spot prices which are different in Region 1 and Region 2. A planned link between the Northern part of NSW and Queensland will allow to transfer 1000 MW to NSW during off-peak hours and about 500 MW to Queensland during high demand periods, see Fig. 8 [1].

A potential undersea DC link between Victoria (VIC) and Tasmania (TAS) will allow the transfer of 300 MW in both directions.

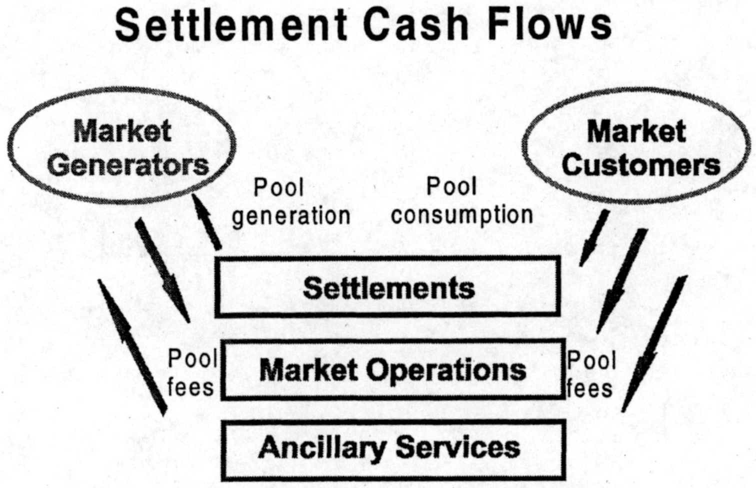


Fig. 7 Structure of settlement process

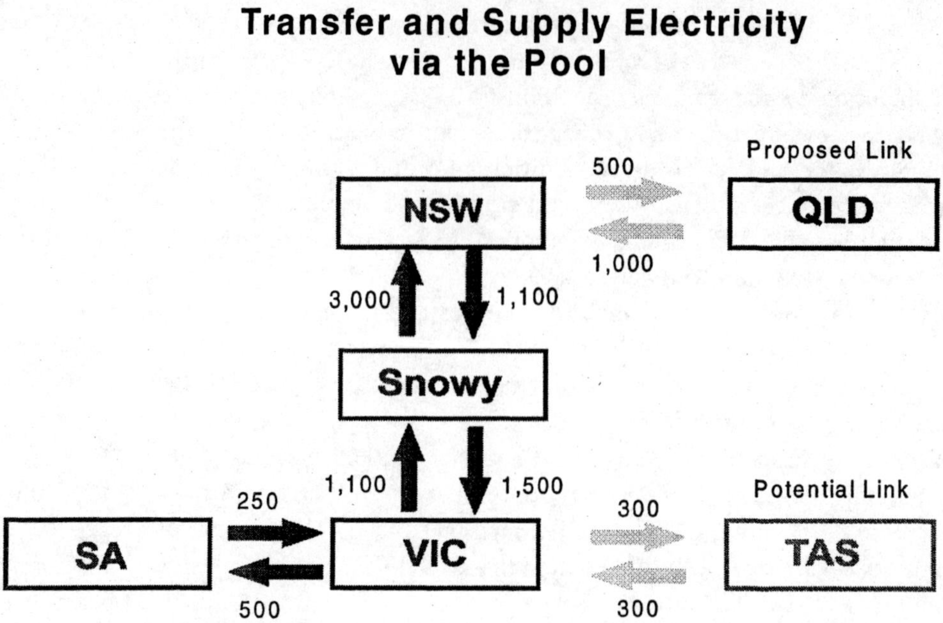


Fig. 8 Flow of electrical energy between regions
Limited capacity of inter-regional links leads to the differences in spot prices, as illustrated by example in Fig. 9.

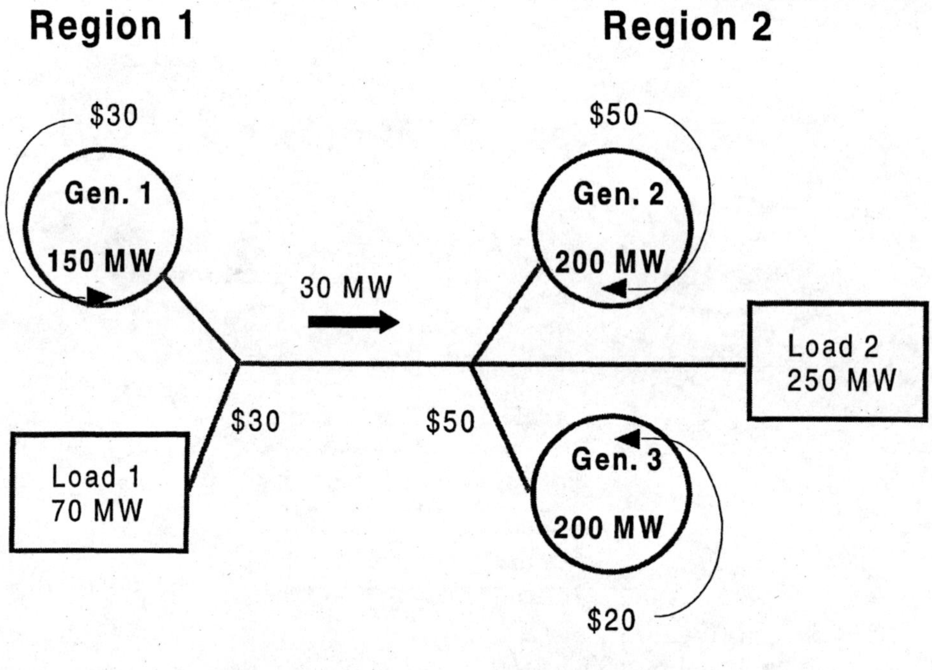


Fig. 9 Impact of limited inter-regional links' capacity on spot prices

In Region 1, there is an excess of electrical energy since Gen. 1 can cover the local demand of 70 MW and export 80 MW to Region 2. In this region, Load 2 requires 250 MW that may be partly covered by 200 MW from Gen. 3 offering electrical energy for 20$/MWh. Additionally, 30 MW can be exported from Region 1 with the price $30. However, Gen. 2 has to deliver additional 20 MW to meet demand from Load 2, therefore this generator sets up a spot price in Region 2 at $50. If the link capacity increases to 50 MW or more, the export from Region 1 would reduce a spot price in Region 2 to $30.

## 7 INTER AND INTRA REGIONAL LOSSES

The energy losses occurred during energy transfer are calculated applying an Inter Regional Loss Factor, as seen in Fig. 10. Each region has its own reference node defined as the physical location of substation from which loss factor is calculated. For example, in NSW, the reference node is located in the substation in Southern suburb of Sydney.

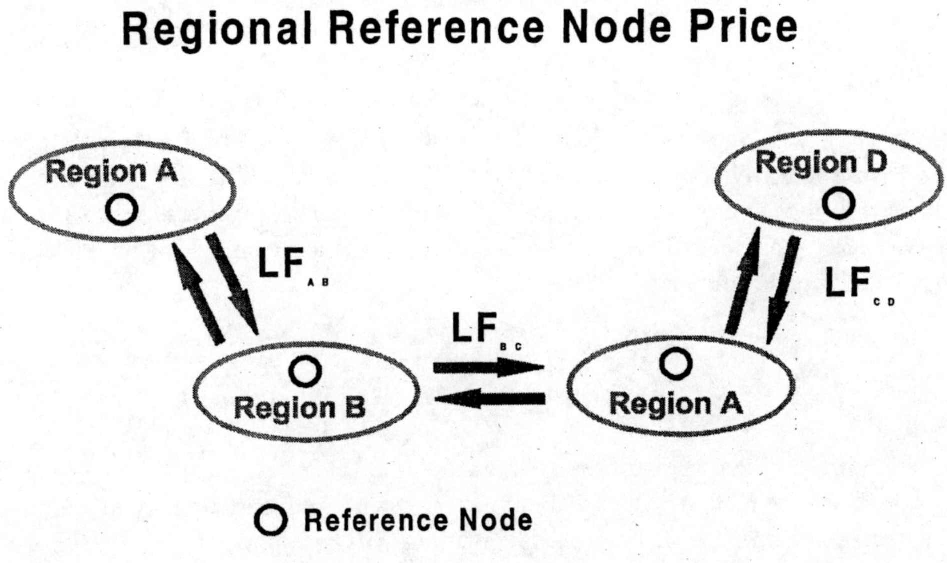


Fig. 10 Inter Regional Loss Factors

Electrical losses within regions are calculated using Intra Regional Loss Factors. The values of Loss Factors are pre-determined according to historical analysis of the electrical losses within regions [1]. In the central dispatch process, offer and bid prices are adjusted by a Intra Regional Loss Factor. This is illustrated in Fig. 11, where Generator offers electrical energy for $50/MWh. The price calculated for him includes intra regional losses, and it is reduced to $48.5 = $50 × 0.97. A spot price in the reference node is equal to $50. A user that is located in some distance from the reference node has Loss Factor equal to 0.3. A price he has to pay for electrical energy is adjusted due to the value of its Loss Factor, that is, \$5.13 = \$50 × 1.03.

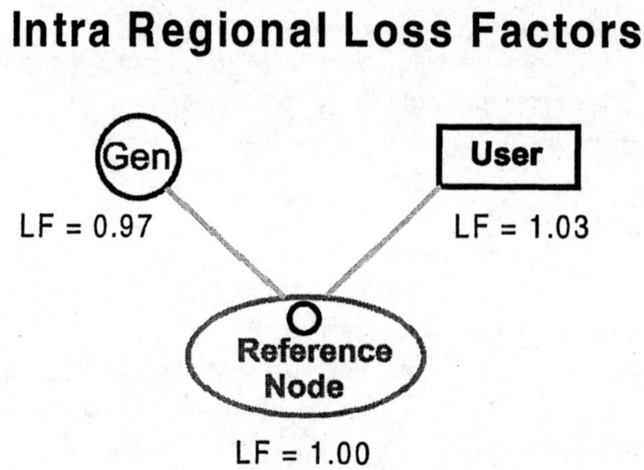


Fig. 11 Price adjustment by Loss Factors

The Loss Factor values are based on historical records of load flow. However, actual losses are changing dynamically with demand. This may lead to differences between the real losses in the pool and the losses calculated by fixed loss factors. When the real losses are larger than calculated, the Market Participants are charged for the difference proportionally to the amount of their energy consumption.

## 8 Regulations

Traditionally, State Governments regulate the electricity supply industry in Australia. When the National Electricity Market is established, participating jurisdictions have agreed that a mixture of state based regulations will be applied [1] [2], see Fig. 12.

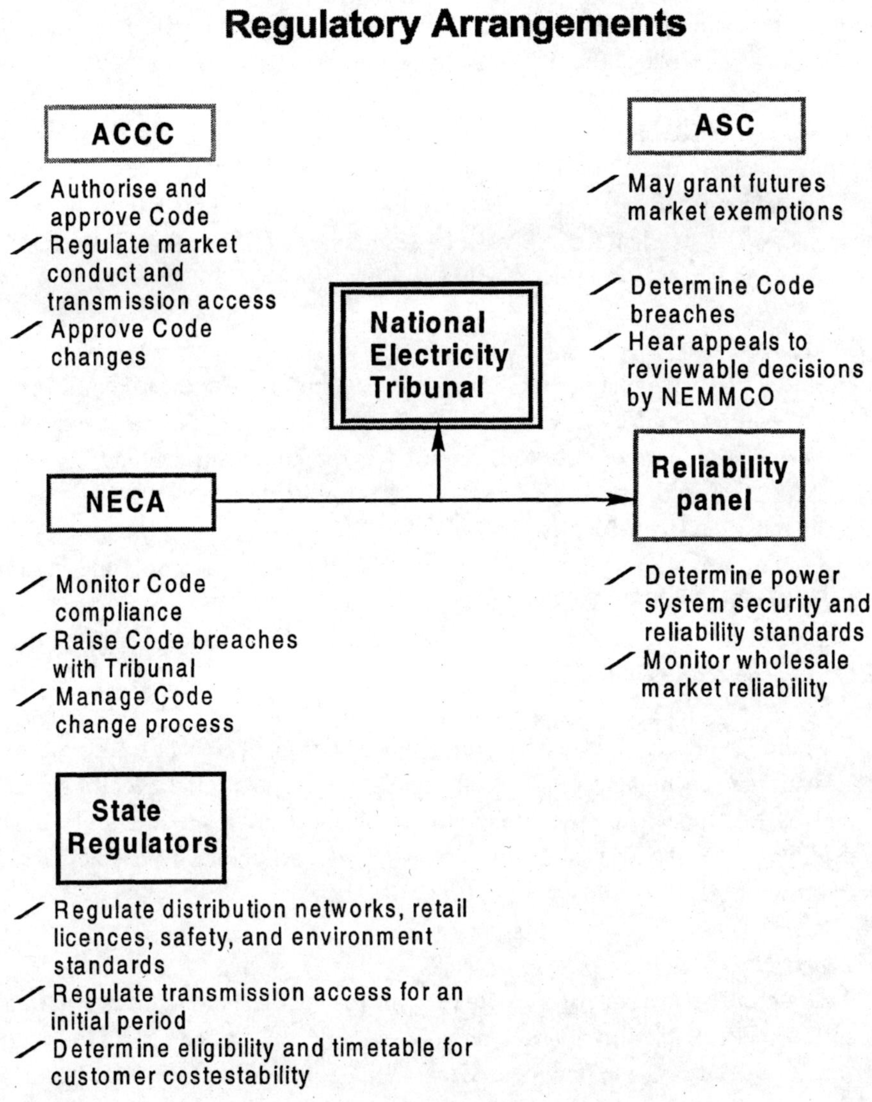


Fig. 12 National Electricity Market regulatory arrangement [1]

The roles of regulators are as follows [1]

***Australian Competition and Consumer Commission (ACCC)***

The ACCC will have a significant role to play in the NEM. Its primary roles will be:

- authorisation of the Code under Part VII of the Trade Practices Act and acceptance of the access undertakings to be provided by Network Services Providers to the ACCC under Part IIIA of the Trade Practices Act;
- authorisation of any changes to the Code;
- an ongoing role in regulating network connection and network access pricing

- ongoing supervision of market conduct and behaviour under Part IV of the Trade Practices Act – particularly in relation to issues of abuse of market power.

***Australian Securities Commission (ASC)***

The ASC has a role in the NEM in that it will determine whether an exempt futures market declaration will be granted to Market Participants under section 1127 of the Corporations Law. This declaration will be fundamental to the ability of Market Participants to trade in hedge contracts for electricity.

***National Electricity Law***

The National Electricity Law supports the effective operation of the Code and the NEM and enables NEMMCO and NECA to fulfil their respective roles in each participating jurisdiction. The National Electricity Law requires, among other things that:

- only NEMMCO may operate a wholesale market for electricity
- generators must register with NEMMCO
- retailers and customers may only purchase electricity from the wholesale market for electricity operated by NEMMCO
- Network Service Providers must register with NEMMCO and agree to be bound by the network access rules in the Code.

The National Electricity Law is cooperative legislation, which has been passed by the South Australian Parliament as the lead legislator. The National Electricity Law is then enacted by other participating jurisdictions. Changes to a jurisdiction's enabling legislation will require the unanimous approval of all participating jurisdictions. This will help to ensure regulatory consistency across the NEM.

***National Electricity Code***

The Code defines the terms of participation in the wholesale electricity market for generators, transmission and distribution network owners and service providers, retailers, and customers. Specific Code chapters deal with: registration of participants with NEMMCO, rules for the operation of the central dispatch process and spot market, security arrangements for the power system, network connection access arrangements including network pricing, metering of Market Participants, and the administration of the Code itself through enforcement, dispute resolution, and a Code change process. The Code also includes a chapter of Code derogation where each participating jurisdiction has specified transitional provisions that will operate to ensure a smooth transition from the existing electricity industry regulatory arrangements to the NEM. Most of these derogations will terminate on or before 31 December 2002.

***National Electricity Tribunal***

Under the National Electricity Law, a Tribunal will be formed to:

- review any decision of NEMMCO and NECA identified within the Code as a reviewable decision
- hear and determine an application by NECA alleging that a Code Participant ahs breached the Code.

*National Electricity Code Administrator Limited (NECA)*

NECA is a company formed by participating jurisdictions to administer the Code. It has a similar corporate structure as NEMMCO and is also controlled by the participating jurisdictions as members. NECA is responsible for:

- monitoring and reporting on Code compliance
- enforcing the Code through the National Electricity Tribunal
- managing changes to the Code
- granting derogation from the Code
- registering Metering Providers in accordance with the Code
- providing a means of effective dispute resolution where other dispute mechanisms created by Code Participants fail.

***NECA's Reliability Panel***

NECA has established a Reliability Panel to monitor, review, and report on the performance of the NEM in terms of power system security and reliability. The panel will, on advice from NEMMCO, and in public consultations, establish the power system security and reliability standards to be followed by NEMMCO as part of the Code. The panel will:

- monitor, review, and report on the performance of the NEM in terms of the reliability of the power system
- determine, on the advice of NEMMCO, reserve margins to handle problems in the power system due to a generating unit or transmission network failure
- determine guidelines for NEMMCO's powers to intervener in the NEM to maintain or re-establish the reliability of the power system.

***State laws and regulation***

Each participating State and Territory will:

- regulate network pricing for transmission networks for transitional periods terminating on or before 31 December 2002, after which regulation will transfer to the ACCC
- regulate network pricing and access for distribution networks
- determine when customers become eligible to participate in the wholesale electricity market defined in the Code or in the retail supply market where contestable customers may trade with the retailer of their choice
- set distribution and retail licence conditions to regulate the behaviour of distributors and retailers supplying electricity to customers

- continue to enforce previously established safety and environmental standards.

## 9 Conclusions

The success of the Victorian electricity market indicates a number of possible benefits from the introduction of the National Electricity Market. They include [1] [4]:

- Reduction of generating resources, since they can be shared across the States
- Deferral in new plant construction due to better generation utilization
- Provision o cheaper electricity from regions with access to low cost energy sources
- More efficient use of electricity by customers

The first stage of the National Electricity Market commenced in May 1997 as NEM1, while the second stage is introduced in May 1998.

## References

1 "*Australia's National Electricity Market – An Introduction*," National Electricity Market Management Company Ltd. November 1997.
2 Van der Mye, S., "Status of Australian Electricity Reform," Econ 98, *Electrical Energy Conference*, March 10-11 1998, Melbourne
3 Spadling, B., "How the Pool Settles," Econ 98, *Electrical Energy Conference*, March 10-11 1998, Melbourne
4 NEM1, *Gateway to the National Electricity Market,* Implementation Edition, May 1997

# Index

## M



## N



## O



## P



## R



## S



## T



## U

## V



## W